1800 1900

CENTENNIAL EDITION

(SECOND EDITION)

TINSMITHS' TOOLS AND MACHINES

..OF THE..

HIGHEST GRADE OF EXCELLENCE

MANUFACTURED BY THE

PECK, STOW AND WILCOX CO.

NEW YORK . CONNECTICUT . OHIO

THE ASTRAGAL PRESS
Mendham, New Jersey

This is a reprint of the original Centennial Edition
of the Peck, Stow & Wilcox Company's
Catalog issued in 1899.

Library of Congress Catalog Card Number 93-71230
International Standard Book Number: 1-879335-38-7

Published by
THE ASTRAGAL PRESS
5 Cold Hill Road
P.O. Box 239
Mendham, New Jersey 07945-0239

Printed in the United States of America

1800 CENTENNIAL. 1900

This Edition of our Catalogue is a reprint of our **Centennial Edition,** published in 1898. It is so arranged that the pages and list prices correspond with those of that issue, as well as with those of our General Catalogue, published during the same year.

This Catalogue represents and describes the various kinds and styles of **Tinsmiths' Tools** and **Machines** which we manufacture and which are now in general use by workers in **Sheet Iron** and **Metals** in this and other countries.

We have endeavored to be **accurate** in our **statements** and the purchaser may rely upon any Machine doing the work as represented. It is now—as it always has been—our aim to **produce Goods of the Highest Degree of Excellence.** Our name is a guarantee of Worth and Best Quality; our Reputation has been hard-earned and is the reward of a **constant desire** to merit the **approval** and **good wishes** of all Tinsmiths in whatever land they live.

We remain,

Very truly yours,

The Peck, Stow & Wilcox Co.

NEW YORK, July 1, 1899.

INDEX.

T

V

W

TINSMITHS' TOOLS AND MACHINES.

TIN FOLDING MACHINES.

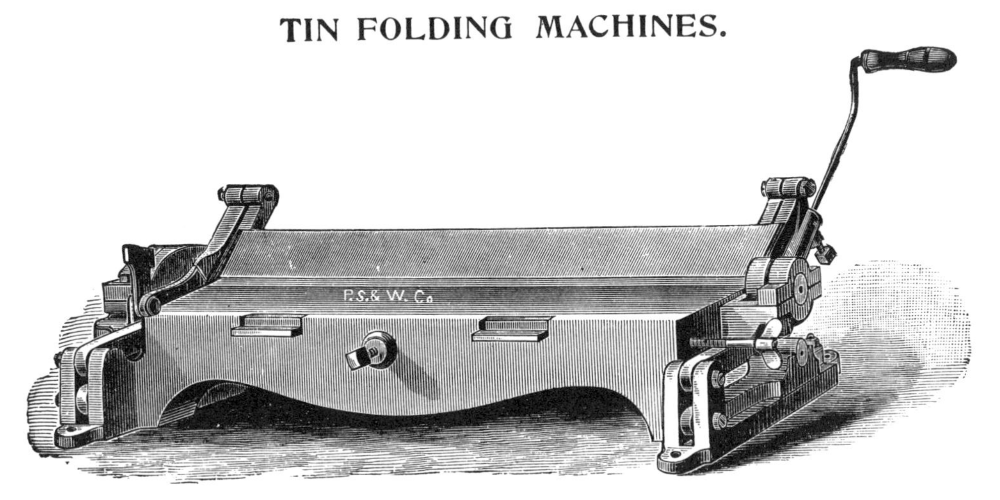

Stow's Patent Adjustable Bar Folders.

WILL TURN LOCKS FROM $\frac{3}{32}$ to $1\frac{5}{8}$ INCHES IN WIDTH.

These machines will form square joints or angles, turn narrow or wide locks, turn a round edge for wiring and form these locks on medium plate with ease. They will also form open or close locks. By open locks we mean such as are suitable for wiring straight work.

Sizes above No. 50 will bend any sheet metal of a thickness not thicker than No. 22 wire guage. When sent from factory these Machines are adjusted properly for common and IX tin plate or other sheet metal of same thickness. If thicker stock is to be used, the machine must be adjusted according to the thickness, by means of the screws at the rear end of the griping jaw bearings.

All the parts are interchangeable; that is, any part of one Machine is exactly like the same part in another. To make it easy to order any piece, all the parts are lettered or numbered, and it is only necessary when ordering new parts to refer to the letter or number on the old part, always giving the number and name of the Machine for which the part is wanted.

The above cut represents the Folding Bar raised at an angle, as in process of folding a sheet of metal.

No. 50.	17 inches, for Tin, will turn Locks from $\frac{3}{32}$ to $\frac{1}{2}$ inch, weighs 50 lbs.,	**$25 00**
No. 52.	20 inches, will turn Locks from $\frac{3}{32}$ to 1 inch, weighs 80 lbs.,	**30 00**
No. 54.	30 inches, will turn Locks from $\frac{3}{32}$ to 1 inch, weighs 155 lbs.,	**40 00**
No. $54\frac{1}{2}$.	36 inches, will turn Close Locks only from $\frac{3}{16}$ to $1\frac{1}{2}$ in., weighs 305 lbs.,	**60 00**
No. 55.	36 inches, will turn Open or Close Locks, $\frac{3}{16}$ to $1\frac{1}{2}$ in., weighs 305 lbs.,	**70 00**
No. 56.	42 inches, will turn Close locks only from $\frac{3}{16}$ to $1\frac{5}{8}$ in., weighs 355 lbs.,	**80 00**
No. 58.	42 inches, will turn Open or Close Locks, $\frac{3}{16}$ to $1\frac{5}{8}$ in., weighs 355 lbs.,	**90 00**

We are also making special Folders for forming any width locks, on light or heavy plate, of any length desired.

TIN FOLDING MACHINES.

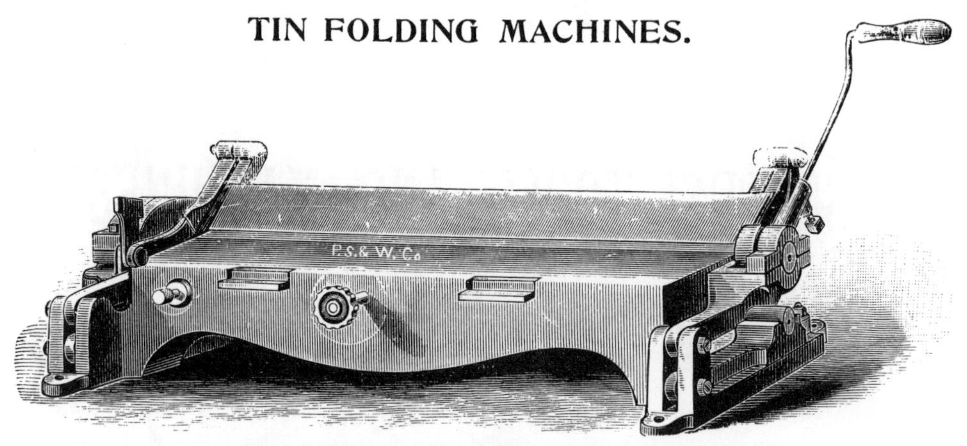

Stow's Patent Adjustable Bar Folders.

WILL TURN LOCKS FROM $\frac{3}{32}$ to 1 INCH IN WIDTH.

These Folders are adapted to the same work as those described on the previous page, though somewhat different in construction. The movement of the gauge is instantaneous. The set screw in front and at the left of the machine holds the gauge firmly in place and secures a fold or lock of uniform width. The parts of these folders are interchangeable and correspond with the parts of our No. 52 Folder.

No. 152. 20¾ inches, will turn Locks from $\frac{3}{32}$ to 1 inch, weighs 80 lbs., . . **$30 00**
No. 154. 31 inches, will turn Locks from $\frac{3}{32}$ to 1 inch, weighs 155 lbs., . . **40 00**

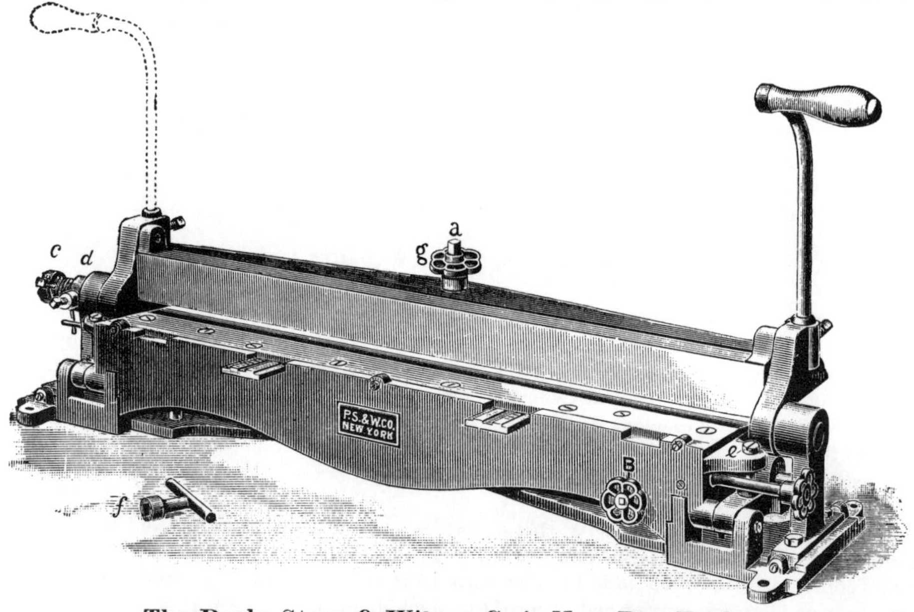

The Peck, Stow & Wilcox Co.'s New Bar Folders.

WILL TURN LOCKS FROM $\frac{3}{32}$ to 1⅝ INCHES IN WIDTH.

We commend these Folders for their real merit and excellence. They embrace the latest mechanical devices and the inventions are secured by *Letters Patent.* Among other advantages we claim that they work easier and will produce a greater variety of work than any other Folders. The wearing parts are more durable and the jaw is steel. The wedge that raises and lowers the wing is operated by a rack and pinion and is, therefore, more easily moved. When set, the wedge can be firmly fastened in its place by the set nut A. When set for turning round edges the wing is flush with the jaw when opened, so that the metal can be readily slipped under the plate, and is automatically released when closed. It has an adjustable stop which can be arranged to hold the folding bar at any desired angle, and the gauge is graduated and moved by a positive mechanism which makes it always turn a true lock; and it can be fastened at any given point by means of the screw B. The folding bar is provided with a spring which can be adjusted to overcome the weight of the bar. All acting parts are adjustable to compensate for wear. Duplicate parts can be furnished.

No. 102. 21 inches, will turn Locks from $\frac{3}{32}$ to 1 inch, weighs 85 lbs., . . **$30 00**
No. 104. 31½ inches, will turn Locks from $\frac{3}{32}$ to 1 inch, weighs 160 lbs., . . **40 00**
No. 105. 37 inches, will turn Locks from $\frac{3}{16}$ to 1⅝ inches, weighs 305 lbs., . . **70 00**
No. 106. 42½ inches, will turn Locks from $\frac{3}{16}$ to 1⅝ inches, weighs 365 lbs., . . **90 00**

TIN FOLDING MACHINES.

WITH SQUARE BOX AND SQUARE PIPE ATTACHMENT.

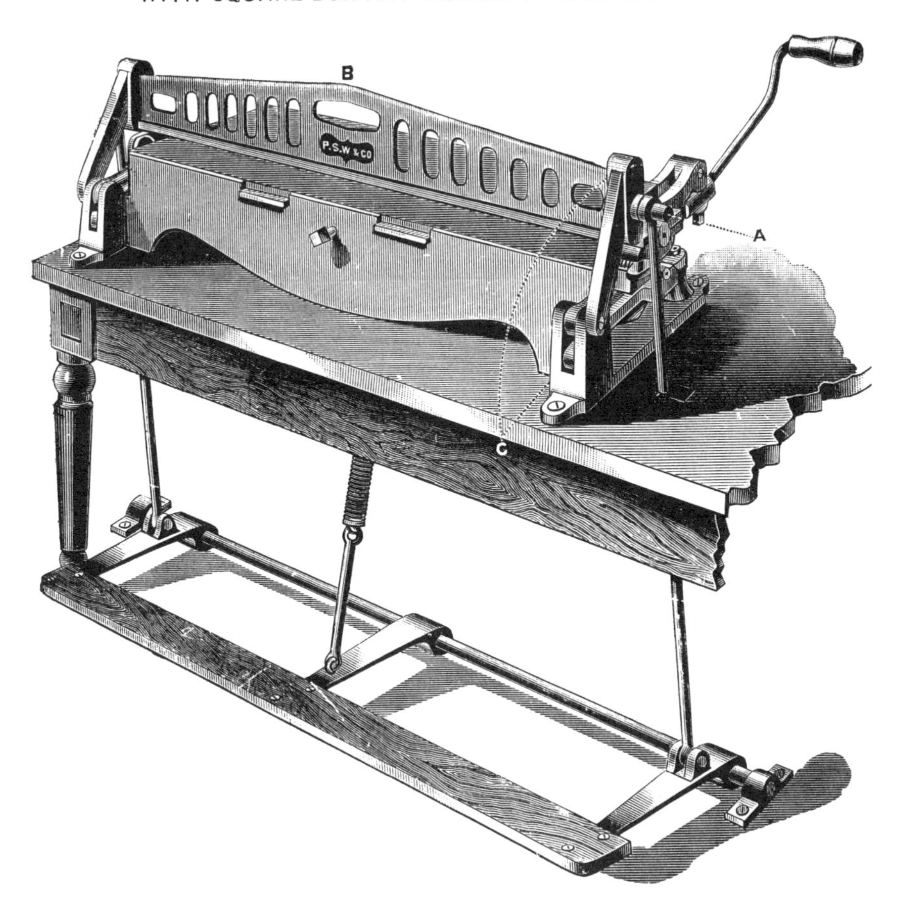

Folder with Square Box Attachment.

We are making our Folders with an attachment by which sharp bends can be made in sheet metal 42 inches wide or less, and at any required distance from either end. These bends can be made at any angle of not more than 55 degrees. This attachment, when not wanted, can be dropped back out of the way.

The above cut represents our regular No. 54 Folder with the square box attachment, on a bench ready for use. In case the Folder only is to be used the attachment B can be turned on the lines marked C and rest out of the way on the bench.

With the No. 54 or 154 Folder an edge can be turned as wide as one inch, and any intermediate width as narrow as 3-32 inch. With the attachment, square pipe can be formed of any size usually required, in sections 30 inches long or less.

No. 54.	Folder with Square Box Attachment, weighs 236 lbs., . .	**$50 00**
No. 154.	Folder with Square Box Attachment, weighs 236 lbs., . .	**50 00**
No. 56.	Folder with Square Box Attachment, weighs 455 lbs., . .	**92 00**

TIN FOLDING MACHINES.

No. 48. Folder, with Foot Lever.

The above cut represents a small Folder, 8 inches long, designed to turn edges very rapidly, for small cans or boxes. This Folder can be made to order to turn any ordinary lock of one width, from 1/8 to 1/2 of an inch. The gauge is not adjustable, and turns a narrow lock for the purpose above named only. It is so constructed as to be used either by hand or foot power. It is very rapid in its execution, and is admirably adapted for the purpose. It is made of steel parts, and is very durable.

No. 47 is the same as No. 48, except it is made with an adjustable gauge and will turn locks of different widths from 3-32 to 1 inch.

No. 47.	Folder, with Adjustable Gauge and Foot Lever, weighs 60 pounds, .	**$35 00**
No. 48.	Folder, with Foot Lever, weighs 60 pounds, ...	**30 00**
No. 48.	Folder, without Foot Lever, weighs 50 pounds, ...	**25 00**

SPECIAL FOLDING MACHINES.

We make a large assortment of Folders of every description for bending sheet metal in various forms and shapes; we also make various attachments for both regular and special folders to meet our customers' demands. Some of these Folders will be found useful for the following purposes: One for turning an 1/8 inch lock or wider, on No. 20 sheet metal, or lighter, 30 inches long, or less; it will also bend a sheet of metal at any point or angle that is 30 inches square or less, and is a convenient machine for cornice and refrigerator work, etc.

We also make another Folder for making Seidlitz Powder Box Lids, and all other similar work out of light sheet metal.

Prices given on receipt of sample or description of work wanted.

We call attention to the following:

No. 49.	14 inches, will turn only one width of lock, but can be furnished to form locks of different widths, weighs 60 pounds, ...	**$24 00**
No. 49½.	14 inches, the same Folder with Adjustable Gauge to turn locks from 3/32 to 1 inch on No. 22 Iron, or lighter; weighs 60 pounds, . .	**30 00**
No. 60.	51 inches, will turn only one width of lock, but can be furnished to turn locks of different widths on No. 18 Iron. It has a simple bar; turns sharp edges only, will not turn round edges. The bar does not drop. This Folder is made in the most substantial manner. Weighs 1050 lbs,	**150 00**
No. 61.	51 inches, the same Folder with Adjustable Gauge turns locks from 1/4 to 2½ inches in width. Gauges move independently of each other, and are graduated to be set alike easily. Handle on each end of machine. Weighs 1050 pounds, ...	**175 00**
No. 80.	85 inches, will turn a close lock on No. 26 Iron. It is made in the same manner as the No. 60 Folder. Weighs 1200 pounds, . .	**225 00**
No. 81.	85 inches, the same Folder with Adjustable Gauge turns locks from 1/4 to 2 inches in width. Weighs 1200 pounds, ...	**250 00**

GROOVING MACHINES.

Stow's Encased Groovers.

No. 1. For Heavy Work, 20 inches, with Rotary Stand, weighs 76 pounds,		**$13 50**
No. 2. For Common Work, 17 inches, with Rotary Stand, weighs 64 pounds,		**11 00**
Extra Stands for either No. 1 or No. 2,	each,	**75**
Extra Rollers (three rolls constitute a set), grooves 1/8, 5/32, 7/32 inch,	each,	**75**
Groover Ratchet, with Brass Mounting,		**2 50**
Groover Pinion,		**50**

Patent Brass Mounted Groovers.

WITH STEEL BAR AND MILLED HORN.

No. 01. For Heavy Work, 20 inches, with stand, weighs 78 pounds,		**$13 50**
No. 02. For Common Work, 17 inches, with stand, weighs 69 pounds,		**11 00**
Extra Stands for either No. 1 or No. 2,	each,	**75**
Extra Rollers (three rolls constitute a set), grooves 1/8, 5/32, 7/32 inch,	each,	**75**
Groover Ratchet, with Brass Mounting.		**3 25**
Groover Ratchet, without Brass Mounting,		**2 25**

SPECIAL GROOVERS.—We make Special Groovers to close side seams on cylinders of all lengths and thicknesses of sheet metal, to be run by hand or power. We can construct them to locate seam on outside or inside of vessel, as desired. Prices given on receipt of sample or description of work.

GROOVING MACHINES.

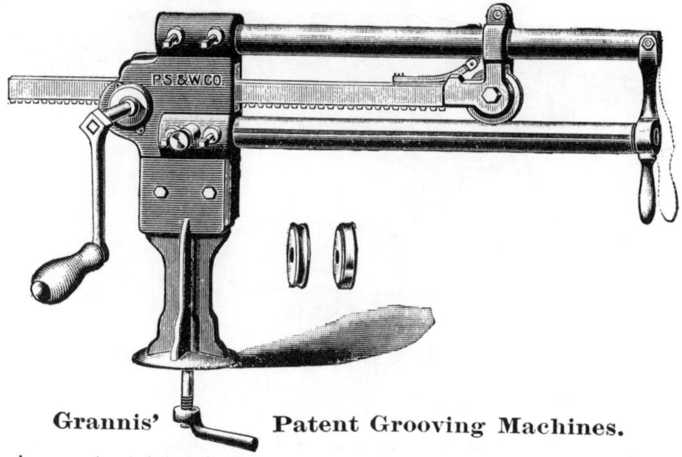

Grannis' Patent Grooving Machines.

This Groover is so constructed that the operator can observe his work during the whole process of grooving. Seams can be formed on either the inside or outside of the cylinder, and it can be used on all vessels not less than two inches in diameter. Four rolls accompany each machine, one each with groove as stated below and one flattening roll. These machines will groove metal not heavier than No. 20 gauge.

No. 11. Grannis' Groover, 20 in., Rolls $\frac{1}{8}$, $\frac{5}{32}$ and $\frac{1}{4}$ in. with Stand, weighs 85 lbs., **$13 50**
No. 12. Grannis' Groover, 30 in., Rolls $\frac{5}{32}$, $\frac{1}{4}$ and $\frac{3}{8}$ in. with Stand, weighs 100 lbs., **18 00**

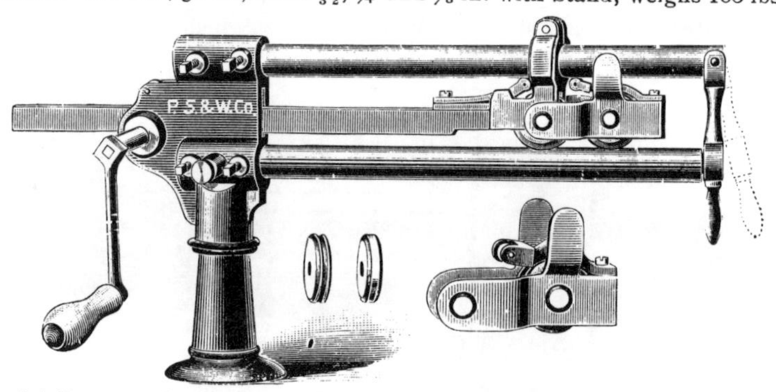

Grannis' Patent Grooving Machines, with Kennedy's Attachment.

No. 21. Grannis' Groover, 20 inches, with Kennedy's Attachment, weighs 90 lbs., **$20 00**
No. 22. Grannis' Groover, 30 inches, with Kennedy's Attachment, weighs 105 lbs., **25 00**

Kennedy's Attachment.

KENNEDY'S PATENT GROOVING ATTACHMENT or FOLLOWING ROLL, for flattening seams, can be used with our regular Brass Mounted Groovers, or with the Grannis' Groovers above; but the same attachment will not work on both machines; hence it will be necessary, in ordering an attachment separate from the machine, to state which machine it is for.

This attachment enables the Tinner to groove and flatten the seam at one operation. The grooving roll is placed in the front attachment, and the flattening or following roll in the place ordinarily occupied by a grooving roll. The grooving roll having passed along the lock and grooved it, the flattening roll follows and flattens it.

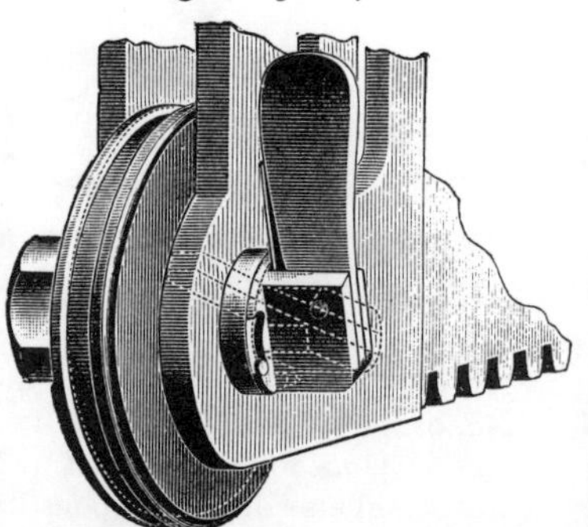

Ridgway's Grooving Roll.

RIDGWAY'S PATENT GROOVING ROLL can be fitted to any Groover. Its method of operation is simple. When the lever is pressed down to a horizontal position it grooves the same as an ordinary grooving roll. Raise the lever as shown in the cut and the roll then flattens the seam.

Kennedy's Patent Grooving Attachment, ... **$7 00**
Ridgway's Patent Grooving Roll, ... **4 50**

GROOVING MACHINES.

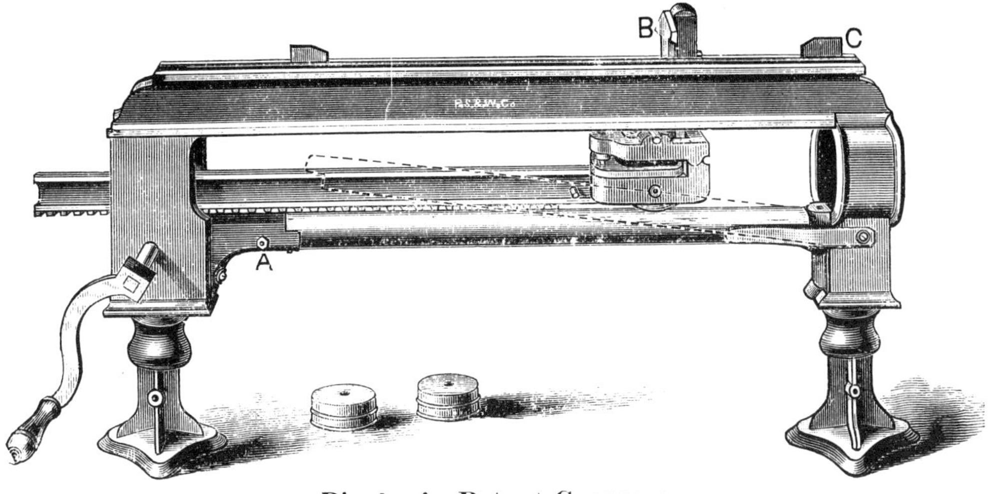

Bigelow's Patent Groovers.

This machine is admirably adapted to both grooving and flattening the seam at one operation. In addition to grooving and finishing stove pipe nicely, all articles made of tin plate can be grooved in this machine by simply changing the rolls. The grooving bar opens automatically at A whenever the grooving roll passes that point, thus permitting all cylinders to be easily removed. C is made to slide in a groove in the upper part of the frame, and whenever B passes that point the grooving roll is changed to a seaming roll, which flattens the seam in the backward motion, thus enabling the operator to adjust the machine for short lengths of work, and avoiding the necessity of running the rolls the entire length of the bar. Three rolls are furnished with these machines, one each for grooving $\frac{11}{32}$, $\frac{13}{32}$, and $\frac{7}{16}$ inch.

No. 1. For Tin and Sheet Iron, 30 inches, complete, with Stands, weighs 180 lbs.,		**$40 00**
No. 2. For Tin and Sheet Iron, 36 inches, complete, with Stands, weighs 220 lbs.,		**50 00**
Extra Rolls, ...	each,	**1 50**
Extra Stands, ...	each,	**1 00**

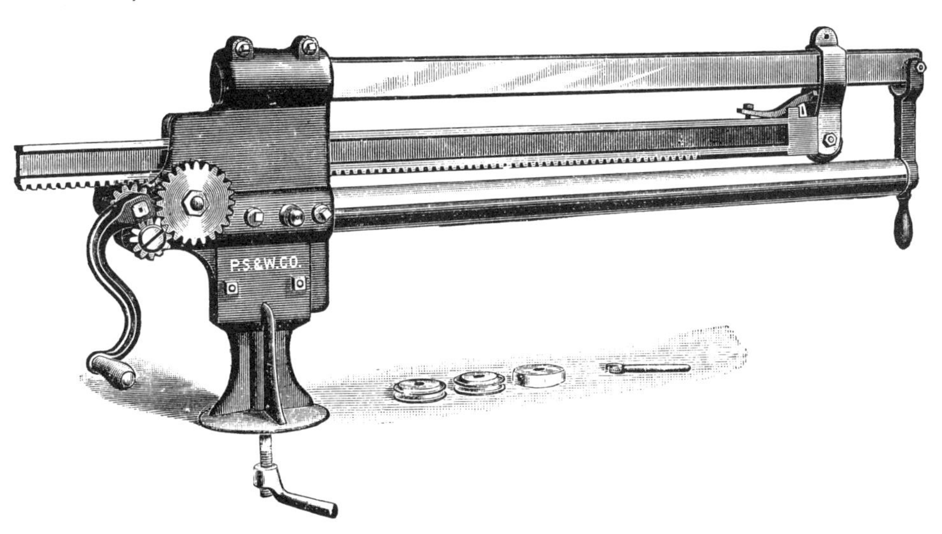

Grannis' Patent Grooving Machine.

This machine is extra heavy and designed for grooving sheet iron as well as tin. The horn is 2½ inches in diameter, is milled for three different widths of seams, and can be easily and quickly adjusted by the operator for the work in hand. It is back-geared 2 to 1 and will groove metal as heavy as No. 22 iron 37 inches in length. Four rolls are furnished with this machine, one each for grooves $\frac{5}{16}$, ⅜ and $\frac{7}{16}$ inch, and one smooth roll.

No. 25. Grannis' Patent Groover, 37 inches, with Stand, weighs 200 lbs., . **$40 00**

STOW'S COLUMBIAN MACHINES.

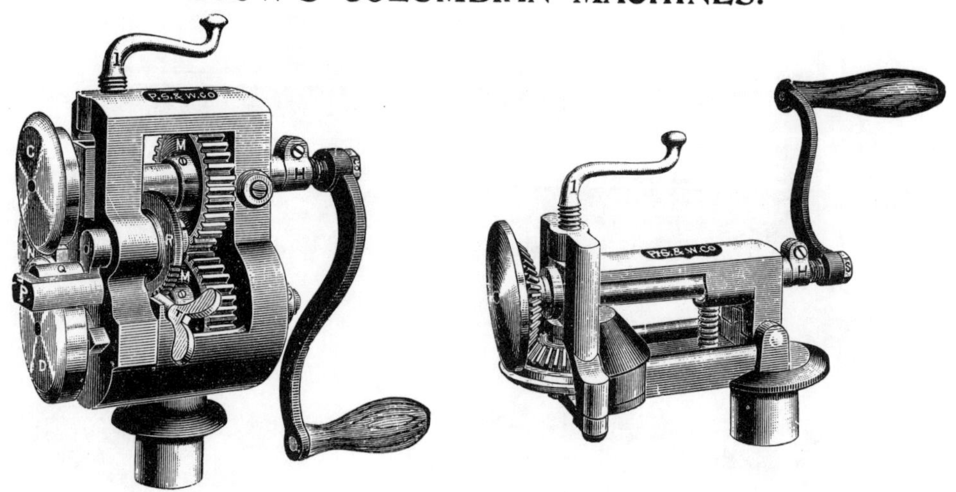

One-fourth size Cuts.

Wiring Machine. **Setting Down Machine.**

Our Columbian Machines are made under recent patents. They have a *solid frame*, are *strong* and *durable*, and are made with *duplicate parts*—that is, the corresponding parts of the same machine are *interchangeable.* By a *patented* arrangement the upper journal boxes are so constructed that the faces readily pass over seams without straining the machine or injuring the work. The Standards used with these Machines are the "Improved," represented on page 29.

Columbian Wiring Machine, with Stand,		**$12 00**
Columbian Wiring Machine, without Stand,		**11 25**
Columbian Setting Down Machine, with Stand,		**9 75**
Columbian Setting Down Machine, without Stand,.		**9 00**
Extra Upper or Lower Roll, with Arbor for Wiring Machine,	each,	**2 50**
Extra Upper or Lower Roll for Setting Down Machine,	each,	**3 00**

One-fourth Size Cuts.

Large Turning Machine. **Small Turning Machine.**

Columbian Large Turning, with Extra Upper and Lower Face, with Stand,		**$10 25**
Columbian Large Turning, with Extra Upper and Lower Face, without Stand,		**9 50**
Columbian Small Turning, with Extra Upper and Lower Face, with Stand,		**10 00**
Columbian Small Turning, with Extra Upper and Lower Face, without Stand,		**9 25**
Extra Faces,	each,	**1 00**

STOW'S COLUMBIAN MACHINES.

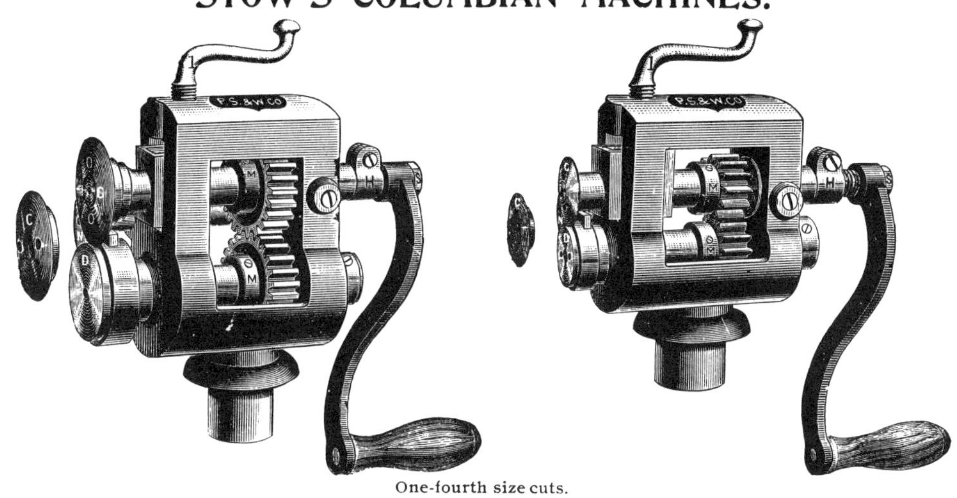

One-fourth size cuts.

Large Burring Machine. **Small Burring Machine.**

Columbian, Large Burr, with Extra Upper Face, with Stand,	$9 00
Columbian, Large Burr, with Extra Upper Face, without Stand,	8 25
Columbian, Small Burr, with Extra Upper Face, with Stand,	8 50
Columbian, Small Burr, with Extra Upper Face, without Stand,	7 75
Extra Upper Faces, each,	1 00
Extra Lower Faces, with Arbor, each,	2 00

COLUMBIAN ELBOW EDGING MACHINE.

One-fourth Size Cuts.

Elbow Edging Machine.

The above cut shows the position of an elbow in turning the edge; also, the form of crease made by the machine to enter the corresponding section in completing the elbow.

Elbow Edging Machine.

The above cut shows the position of the pipe or elbow in forming the bead to receive the creased section, as shown in the illustration at the left.

In using this machine, one-third of the time spent in putting elbows together can be saved by the operator.

The rolls or faces are also sold separate from the machine, and can be used with and are readily adapted to either our Columbian or Encased Small Turning Machine.

Columbian Elbow Edging Machine, with Stand,	$11 50
Elbow Edging Rolls or Faces, separate from Machine, per set (1 pair),	2 50

STOW'S ENCASED MACHINES.

WITH ADJUSTABLE BOXES AND DUPLICATE PARTS.

One-third Size Cut.

Wiring Machine.

Encased Wiring Machine, with Standard,		$14 00
Encased Wiring Machine, without Standard,		13 25
Extra Upper or Lower Roller, with Arbor,	each,	2 50

Extra Faces for Encased Wiring Machines are only furnished with the Arbor or Shaft.

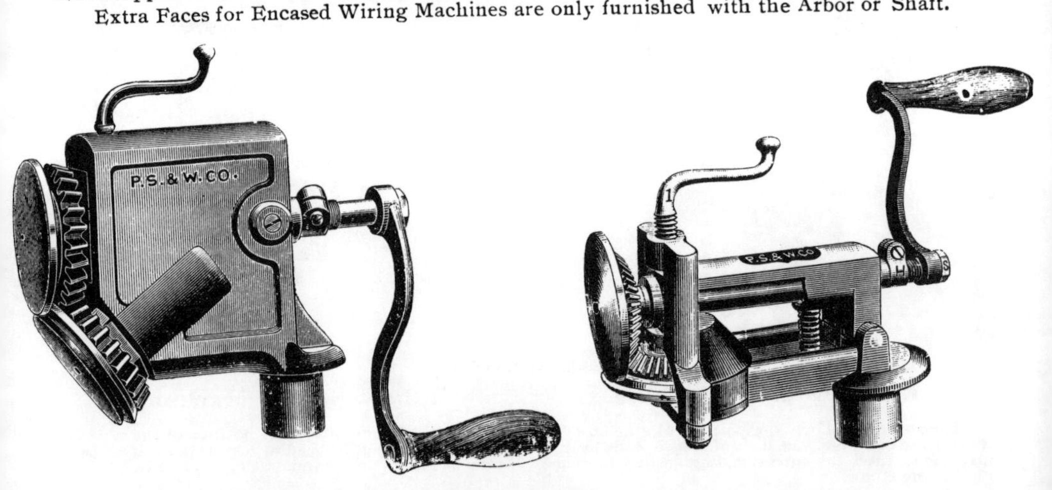

One-fourth Size Cuts.

American Setting Down Machine. **Setting Down Machine.**

The American Setting Down Machine illustrated above has our patented journal boxes so arranged that the faces readily pass over seams without straining the machine or injuring the work. The rolls are nearly parallel, allowing the work to pass through easily. It is exceedingly well made and strong, and can be used on metal as heavy as No. 22 gauge.

Encased Setting Down Machine, with Standard,		$9 75
Encased Setting Down Machine, without Standard,		9 00
Extra Heavy Setting Down Machine, for seams $\frac{3}{16}$ inch wide, with Standard,		11 00
American Setting Down Machine, with Standard,		9 75
American Setting Down Machine, without Standard,		9 00
Extra Upper or Lower Rolls for Setting Down Machine,	each,	3 00
Lower Journal Plate for Encased Setting Down Machine,		50

STOW'S ENCASED MACHINES.

WITH ADJUSTABLE BOXES AND DUPLICATE PARTS.

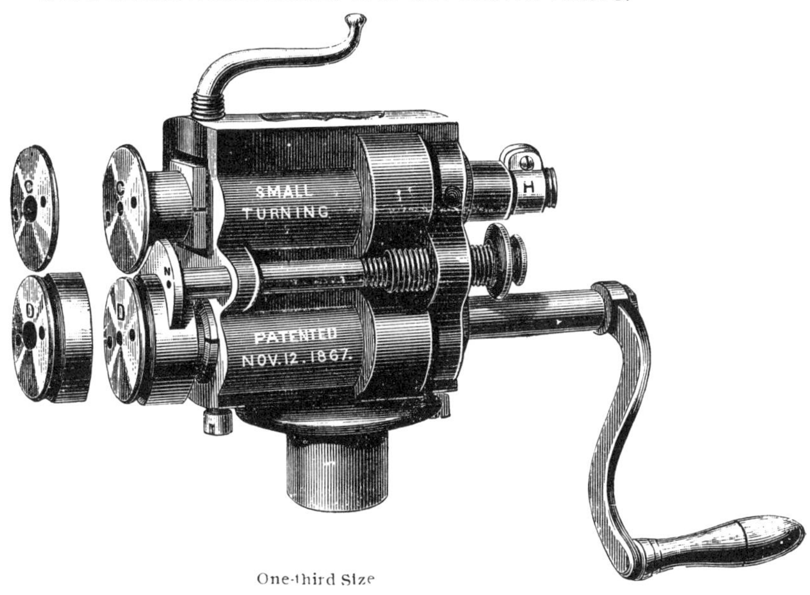

One-third Size

Small Turning Machine.

WITH EXTRA FACES.

Encased Small Turning, with Extra Upper and Lower Face, with Standard, .	$11 25
Encased Small Turning, with Extra Upper and Lower Face, without Standard, .	10 50
Encased Extra Small Turning, same size Faces as Small Burr, with Standard, .	12 00
Extra Faces, ... each,	1 00

One-third Size Cut.

Large Turning Machine.

WITH EXTRA FACES.

Encased Large Turning, with Extra Upper and Lower Face, with Standard, .	$11 50
Encased Large Turning, with Extra Upper and Lower Face, without Standard, .	10 75
Extra Faces, ... each,	1 00

STOW'S ENCASED MACHINES.

WITH ADJUSTABLE BOXES AND DUPLICATE PARTS.

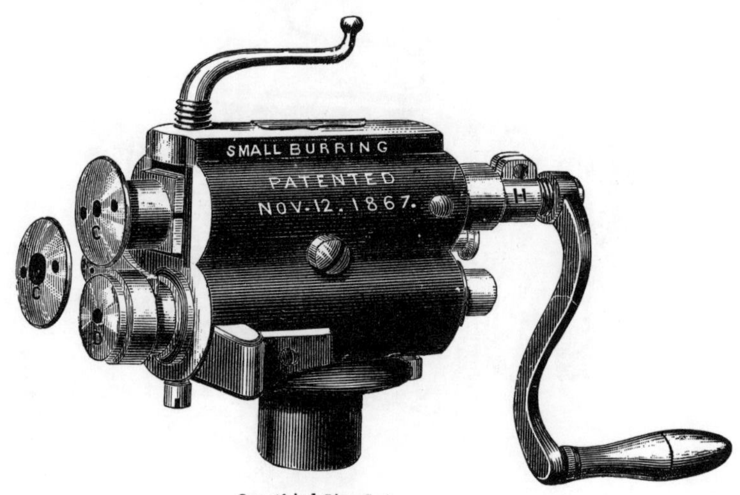

One third Size Cut.

Small Burring Machine.

WITH EXTRA UPPER FACE.

Encased Small Burr, with Extra Upper Face, with Standard, ...		**$10 00**
Encased Small Burr, with Extra Upper Face, without Standard, ...		**9 25**
Extra Upper Faces, ...	each,	**1 00**
Extra Lower Faces, with Arbor, ...	each,	**2 00**

One-third Size Cut.

Large Burring Machine.

WITH EXTRA UPPER FACE.

Encased Large Burr, with Extra Upper Face, with Standard, ...		**$10 50**
Encased Large Burr, with Extra Upper Face, without Standard, ...		**9 75**
Extra Upper Faces, ...	each,	**1 00**
Extra Lower Faces, with Arbor, ...	each,	**2 00**

IMPROVED MACHINE STANDARDS.

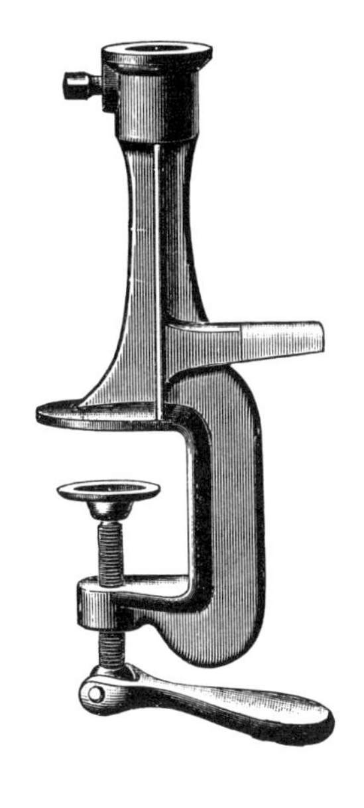

Wiring Machine Standard.

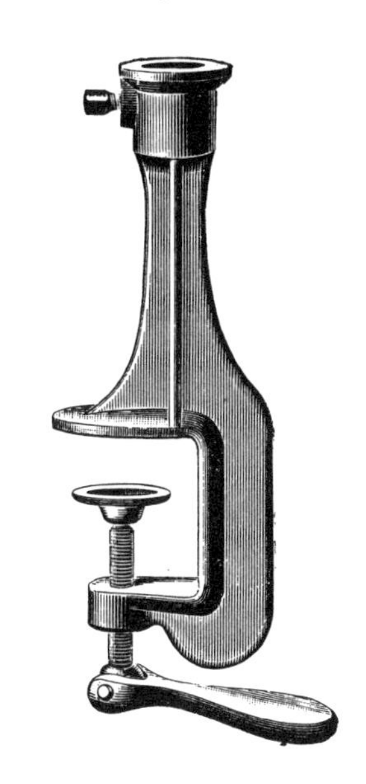

Small Machine Standard.

ENCASED AND COLUMBIAN MACHINES ARE ALL PACKED WITH THESE STANDARDS.

These Standards are so made as to be used on any bench varying in thickness from 1 to 3½ inches. The necessity of cutting holes in the benches is obviated, and the Tinner is enabled to use a Machine in any part of his shop most convenient to his work. The WRENCH IS ALWAYS attached to the Standard.

These Standards are also adapted to RAYMOND'S and No. 1 Machines,

Improved Machine Standards, ...	each,	**$1 00**

A Full Set of Stow's Encased Machines is made up as follows:

Stow's Adjustable Bar Folder, No. 52, 20 inches, ...	**$30 00**
Stow's Encased Grooving Machine, No. 1, 20 inches, with Standard, . .	**13 50**
Stow's Encased Wiring Machine, with Standard, ...	**14 00**
Stow's Encased Setting Down Machine, with Standard, ...	**9 75**
Stow's Encased Large Turning Machine, with Standard, ...	**11 50**
Stow's Encased Small Turning Machine, with Standard, ...	**11 25**
Stow's Encased Large Burring Machine, with Standard, ...	**10 50**
Stow's Encased Small Burring Machine, with Standard, ...	**10 00**
Full Set of Encased Machines, with Standards, ...	**$110 50**
Full Set of Encased Machines, without Standards, ...	**106 00**
Full Set of Encased Machines, without Folder and Groover, ...	**67 00**

RAYMOND'S PATENT MACHINES.

WITH ADJUSTABLE BOXES.

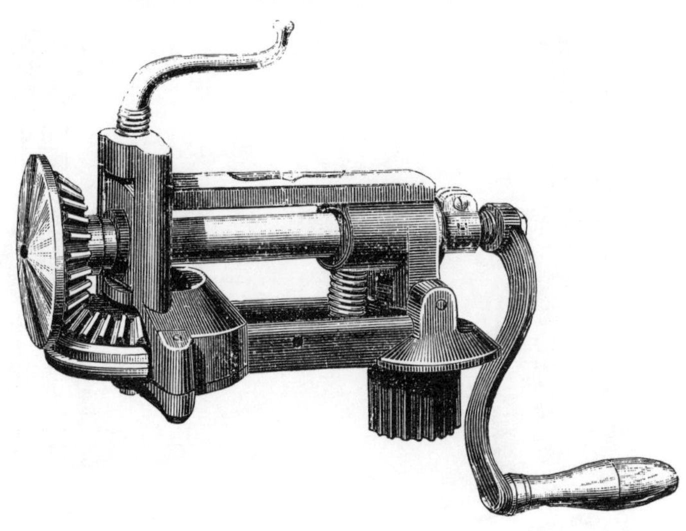

Setting Down Machine.

Raymond's Setting Down Machine, with Stand, ...	$9 75
Raymond's Setting Down Machine, without Stand, ...	9 00
Extra Upper or Lower Rolls for Setting Down Machine, ... each,	3 00
Lower Journal Plate for Raymond's Setting Down Machine, ...	50

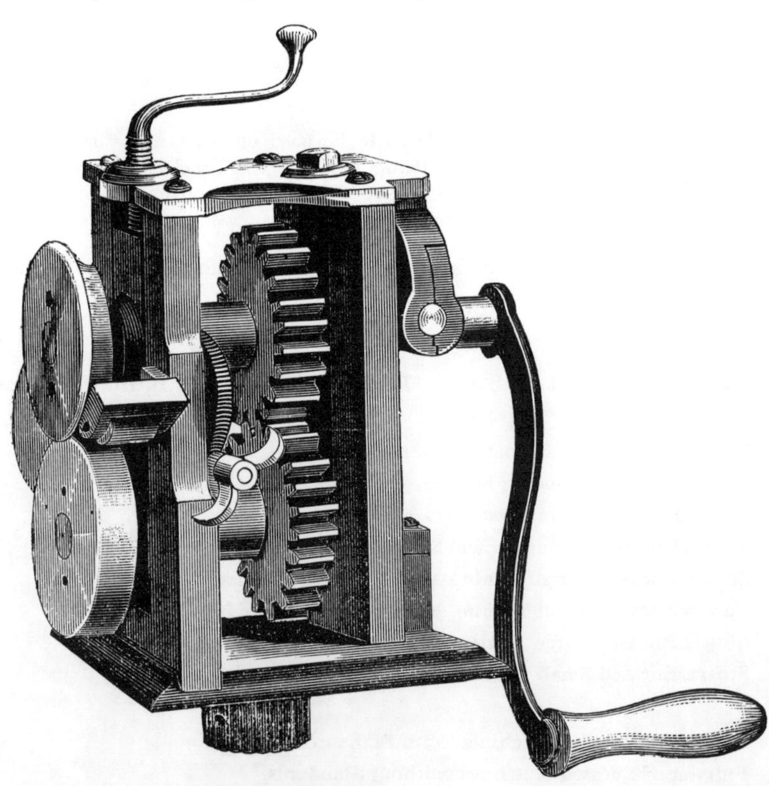

Wiring Machine.

Raymond's Wiring Machine, with Stand, ... ,	$12 00
Raymond's Wiring Machine, without Stand, ...	11 25
Extra Faces for Wiring Machine, ... , . each,	1 25

NOTE.—Improved Standards are adapted to all of Raymond's Machines.

RAYMOND'S PATENT MACHINES.

WITH ADJUSTABLE BOXES.

Small Turning Machine.

Raymond's Small Turning, with Extra Upper and Lower Face, with Stand,	.	**$10 00**
Raymond's Small Turning, with Extra Upper and Lower Face, without Stand,	.	**9 25**
Extra Faces, ...	each,	**1 00**

Large Turning Machine.

Raymond's Large Turning, with Extra Upper and Lower Face, with Stand,	.	**$10 25**
Raymond's Large Turning, with Extra Upper and Lower Face, without Stand,		**9 50**
Extra Faces, ...	each,	**1 00**

RAYMOND'S PATENT MACHINES.

WITH ADJUSTABLE BOXES.

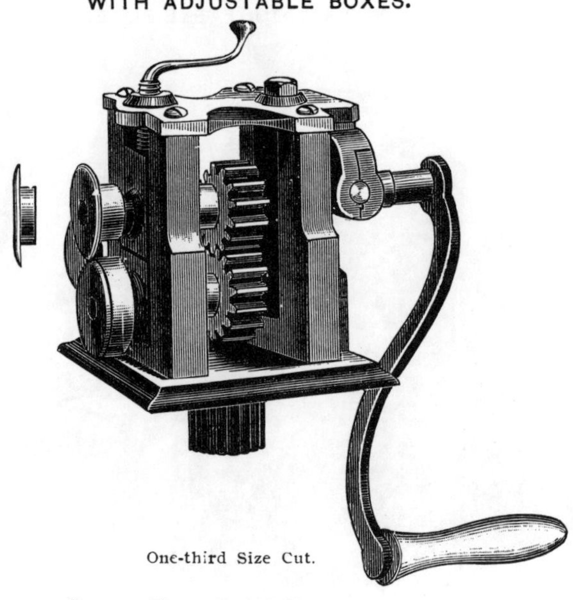

One-third Size Cut.

Small Burring Machine.

Raymond's Small Burr, with Extra Upper Face, with Stand, ...		$8 50
Raymond's Small Burr, with Extra Upper Face, without Stand, . . .		7 75
Extra Upper Faces, ...	each,	1 00
Extra Lower Faces, with arbor, ...	each,	2 00

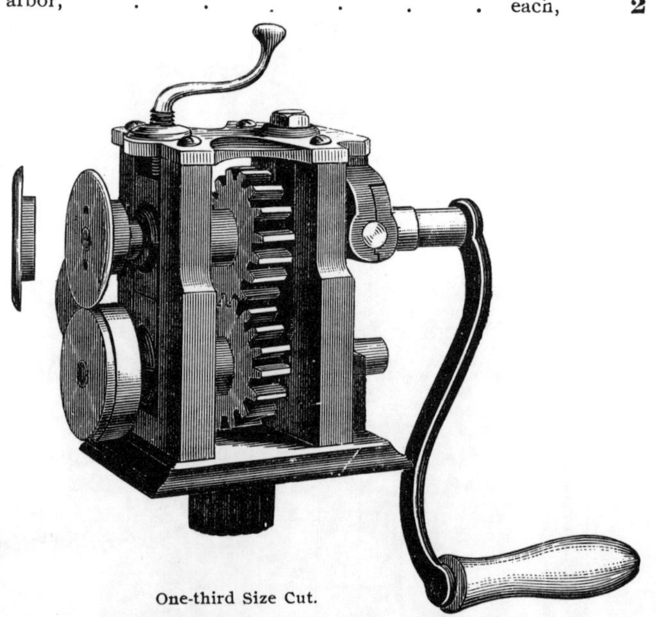

One-third Size Cut.

Large Burring Machine.

Raymond's Large Burr, with Extra Upper Face, with Stand, ...		$9 00
Raymond's Large Burr, with Extra Upper Face, without Stand, . . .		8 25
Extra Upper Faces, ...	each,	1 00
Extra Lower Faces, with arbor, -	each,	2 00

RAYMOND'S MACHINE STANDARDS.

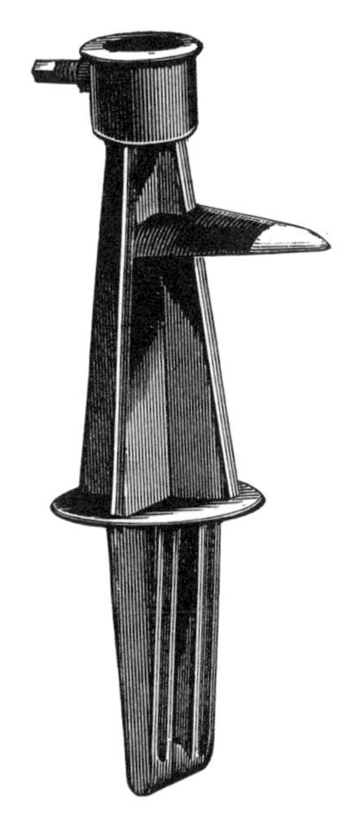

Wiring Machine Standard.

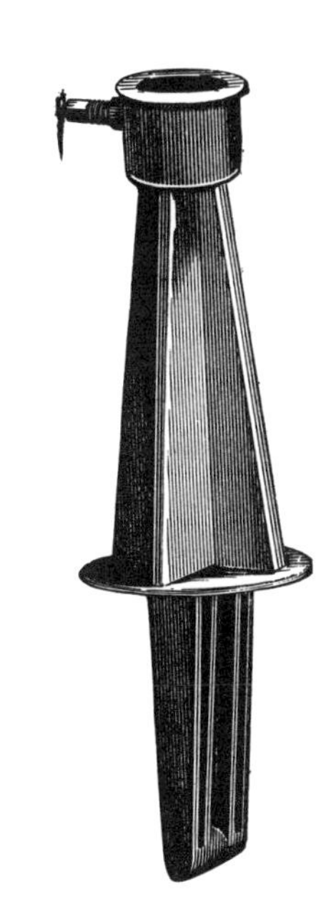

Small Machine Standard.

Raymond's Machines are packed with these Standards, but may be ordered with Improved Standards at an additional cost of 25 cents for each Machine.

Raymond's Machine Standards, ... each, **$0 75**

A Full Set of Raymond's Machines is made up as follows:

Stow's Adjustable Bar Folder, No. 152, 20¾ inches, ...	**$30 00**
Brass Mounted Groover, No. 01, 20 inches, with Stand, ...	**13 50**
Raymond's Wiring Machine, with Stand, ...	**12 00**
Raymond's Setting Down Machine, with Stand, ...	**9 75**
Raymond's Large Turning Machine, with Stand, ...	**10 25**
Raymond's Small Turning Machine, with Stand, ...	**10 00**
Raymond's Large Burring Machine, with Stand, ...	**9 00**
Raymond's Small Burring Machine, with Stand, ...	**8 50**
Full Set of Machines, as above, with Stands, ...	**$103 00**
Full Set of Machines, as above, with Improved Stands, ...	**104 50**
Full Set of Machines, as above, less Folder and Groover, . .	**59 50**

No. 1 MACHINES.

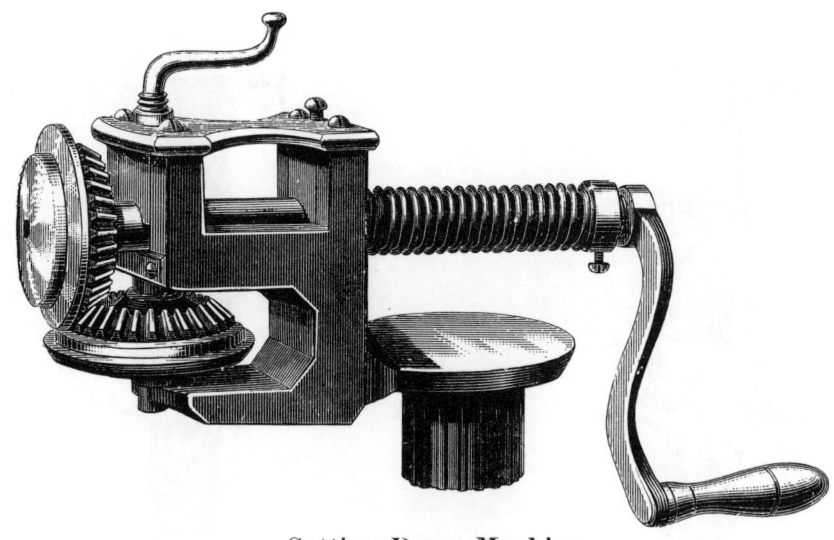

Setting Down Machine.

No. 1. Setting Down Machine, with Stand, ...	$9 75
No. 1. Setting Down Machine, without Stand, ...	9 00
Upper Roll and Arbor for No. 1 Setting Down Machine, . . .	3 00
Lower Roll for No. 1 Setting Down Machine, ...	3 00

Wiring Machine.

No. 1. Wiring Machine, with Stand, ...		$12 00
No. 1. Wiring Machine, without Stand, ...		11 25
Extra Faces for No. 1 Wiring Machine, ...	each,	1 25

NOTE.—Improved Standards are adapted to all No. 1 Machines.

No. 1 MACHINES.

Small Turning Machine.

No. 1.	Small Turning, with Extra Upper and Lower Face, with Stand, .	$10 00
No. 1.	Small Turning, with Extra Upper and Lower Face, without Stand, .	9 25
No. 1.	Extra Small Turning, same size Faces as Extra Small Burr, with Stand,	12 00
Extra Faces,	. . . . each,	1 00

Large Turning Machine.

No. 1.	Large Turning, with Extra Upper and Lower Face, with Stand, .	$10 25
No. 1.	Large Turning, with Extra Upper and Lower Face, without Stand, .	9 50
Extra Faces,	. . . . each,	1 00

No. 1 MACHINES.

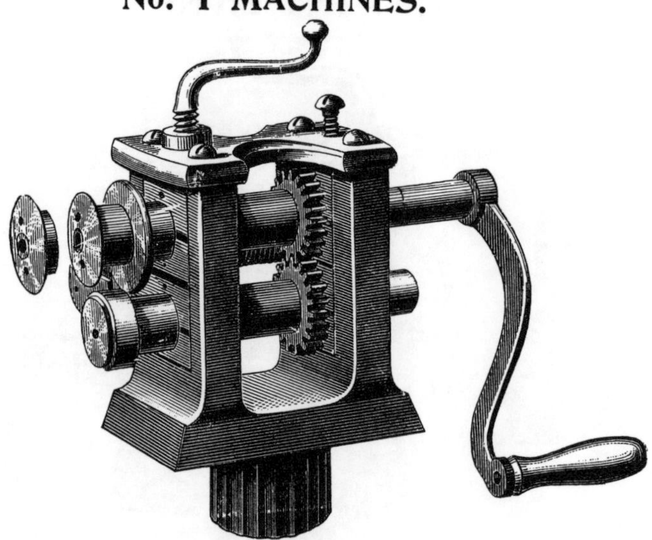

One-third Size Cut.

Small Burring Machine.

No. 1. Small Burr, with Extra Upper Face, with Stand, ...	$8 50
No. 1. Small Burr, with Extra Upper Face, without Stand, ...	7 75
Extra Upper Faces, ... each,	1 00
Extra Lower Faces and Arbor, ... each,	2 00

One third Size Cut.

Large Burring Machine.

No. 1. Large Burr, with Extra Upper Face, with Stand, ...	$9 00
No. 1. Large Burr, with Extra Upper Face, without Stand, ...	8 25
Extra Upper Faces, ... each,	1 00
Extra Lower Faces and Arbor, ... each,	2 00

No. 1 MACHINE STANDARDS.

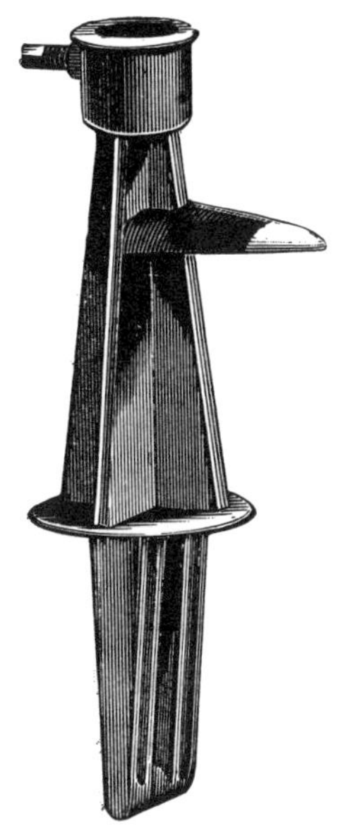

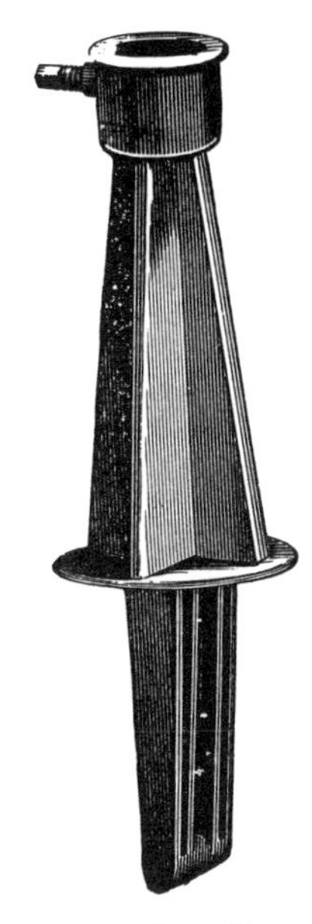

Wiring Machine Standard. **Small Machine Standard.**

No. 1. Machine Standards, ... each, **$0 75**

No. 1 Machines are packed with these Standards, but may be ordered with Improved Standards at an additional cost of 25 cents for each Machine.

A Full Set of No. 1 Machines is made up as follows:

No.	152.	Stow's Adjustable Bar Folder, 20¾ inches, ...	**$30 00**
No.	01.	Brass Mounted Groover, 20 inches, ...	**13 50**
No.	1.	Wiring Machine, with Stand, ...	**12 00**
No.	1.	Setting Down Machine, with Stand, ...	**9 75**
No.	1.	Large Turning Machine, with Stand, ...	**10 25**
No.	1.	Small Turning Machine, with Stand, ...	**10 00**
No.	1.	Large Burring Machine, with Stand, ...	**9 00**
No.	1.	Small Burring Machine, with Stand, ...	**8 50**

Full Set of No. 1 Machines, as above, with Stands, ... **$103 00**

Full Set of No. 1 Machines, as above, with Improved Stands, ... **104 50**

Full Set of No. 1 Machines, as above, less Folder and Groover, . . . **59 50**

STOVE PIPE CRIMPERS AND BEADERS.

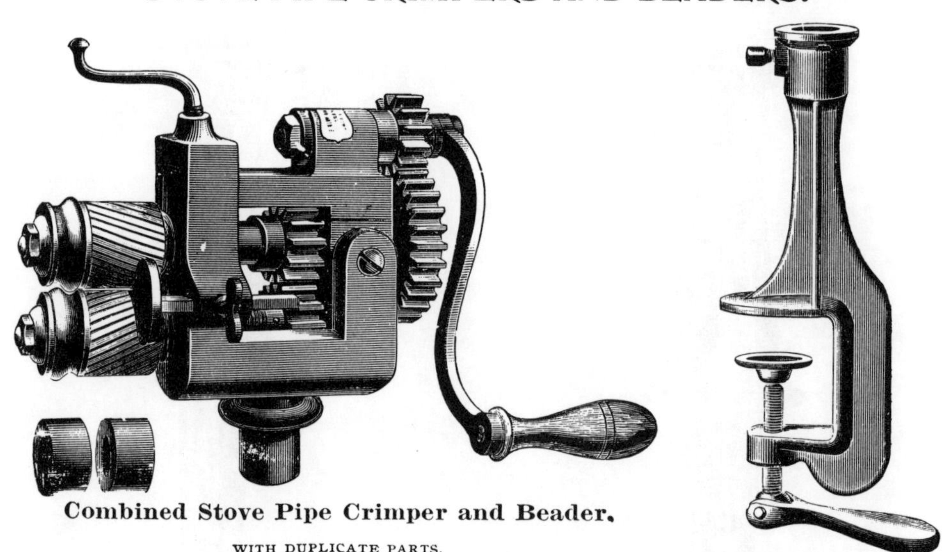

Combined Stove Pipe Crimper and Beader.

WITH DUPLICATE PARTS.

This Machine is designed to facilitate the making and putting together of metal pipe of different diameters. It crimps and contracts the edges of stove and conductor pipe, so that the lengths are put together easily. It is also provided with beading rolls, which bead the pipe at the same time it is crimped. It has steel rolls. When crimping only is desired, substitute the Collars in place of the Ogee Rolls. The gear is made much larger than formerly, and the whole machine is stronger in every way Straight Crimping Rolls can be furnished when desired. This machine can be furnished without back gearing.

We make a Power Crimping Machine for crimping stove pipe. It makes sixty revolutions per minute; pulley, 10 inches; starts and stops by clutch to work by foot.

For description and price of extra parts see page 131.

No. 7.	Stove Pipe Crimper and Beader complete, with Stand, weighs 45 lbs , .	**$12 00**
No. 9.	Stove Pipe Crimper and Beader complete, with Foot Lever, weighs 55 lbs.,	**14 00**
No. 11.	Power Crimping Machine, weighs 150 lbs , ...	**200 00**

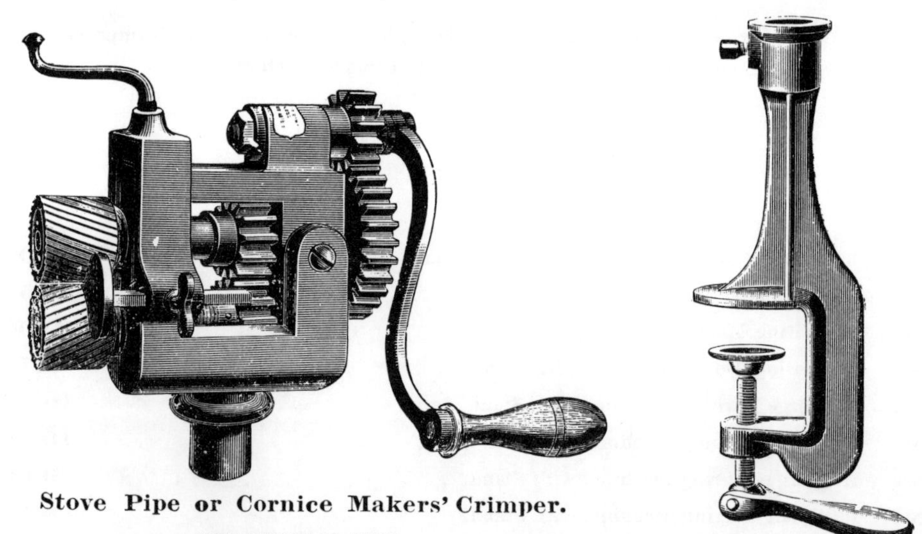

Stove Pipe or Cornice Makers' Crimper.

WITH DUPLICATE PARTS.

This Machine is similar in every respect to the one illustrated above except that it is arranged to be operated without the beader rolls. This allows the machine to crimp close up to a bend or angle. It will be found serviceable for Cornice Work, etc.

No. 17.	Stove Pipe Crimper complete, with Stand, weighs 43 lbs., . .	**$11 00**
No. 19.	Stove Pipe Crimper complete, with Foot Lever, weighs 53 lbs., . .	**13 00**

STOW'S PATENT RIM MACHINE.

One-third Size Cut.

Patent Rim Machine.

This Machine will form, flare and edge straight strips of tin into perfect rims at one operation. It contracts but does not corrugate the metal. It is adjustable, so that rims of different widths can be formed. It is admirably adapted for general use in tin shops, as all strips of metal can be utilized. It has been thoroughly tried and will do the work well.

Stow's Patent Rim Machine, with Stand, weighs 18 lbs., ...		**$17 50**
Extra Upper or Lower Face, ...	each,	**1 50**

TUCKING MACHINE.

One-third Size Cut.

Improved Tucking or Crimping Machine.

This Machine is used by manufacturers of canning boxes, and is very useful for their special work.

No. 1. Improved Tucking or Crimping Machine, with Stand, weighs 13 lbs., . **$12 00**

POWER WIRING AND TURNING MACHINES.

Power Turning Machine.

These Machines are intended for use on very heavy work, and are suitable for Brass Kettles and other vessels made of heavy sheet metal. They can be operated on Sheet Steel as heavy as No. 16 gauge. Wire can be used as large as *one inch* in diameter.

The Turning Machine is used in bending or curving the top of the vessel for the wire, which is inserted to give it additional strength and finish. The Wiring Machine is used to finish the operation, by completely and compactly covering the wire.

Wiring Machine, for Power, Weighs 200 lbs., ... **$125 00**
Large Turning Machine, for Power, weighs 200 lbs,, ... **125 00**

CRIMPING MACHINES.

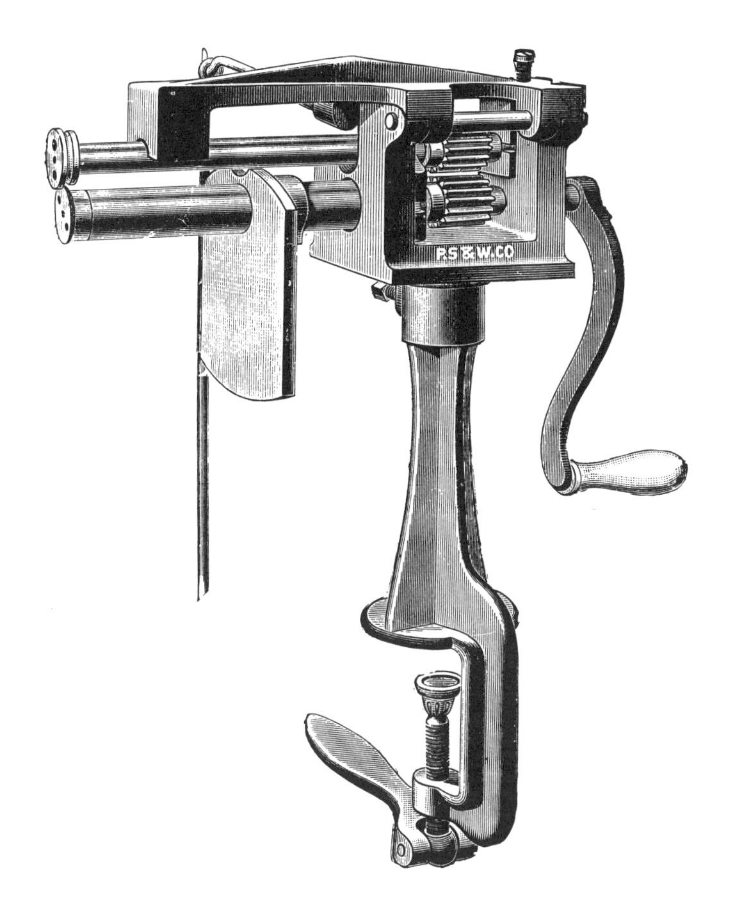

Crimping Machine, with Stand.

No. 1 will put bottoms on all cylinders of a diameter not less than 2½ inches, and in length not exceeding 7½ inches. No. 2 can be used for cylinders of a diameter not less than 1½ inches, and not longer than 6 inches.

The gauges are so arranged as to be adjusted with ease and accuracy.

These Machines can also be arranged to run by power.

No. 1. Crimping, for putting Tops and Bottoms on Boxes, Cans, Cups, etc., with Stand, weighs 68 lbs., ... **$18 00**

No. 2. Crimping, for similar purposes, with Stand, weighs 45 lbs., . . **15 00**

Extra Faces for Crimping Machines, ... each, **1 50**

BEADING MACHINE.

No. 00 Beading Machine, on Stand.

The above cut represents our No. 00 Beading Machine, with Supporting Bar. It is constructed for extra heavy work, and is intended for beading or corrugating very heavy metal. It is backgeared fifteen to one. It will bead or corrugate sheet metal as thick as ⅛ inch. This Machine is made for power, but can be constructed for hand use if desired.

Cutting discs can be applied to this beader, so that it can be used as a circular or slitting shear.

No. 00.Beader complete, with one pair Rolls, weighs 1225 lbs., . .		**$360 00**
Extra Rolls, ...	per pair,	**24 00**
Embossing Rolls, ...	per pair,	**100 00**
Cutting Discs, ...	per pair,	**18 00**

BEADING MACHINE.

No. 0 Beading Machine, on Stand.

The above cut represents our No. O Beading Machine with supporting bar. This Machine will bead iron or soft sheet steel as thick as No. 14 wire gauge, and 10½ inches from the edge of the article to the bead. The Bead shown is "Ogee" and is about 2 inches wide. Other styles or designs can be furnished to order.

In Beading light metal the balance wheel with shaft can be removed and the handle screwed into the large gear, when the Machine will run much more rapidly than with the balance wheel.

This Machine is ordinarily made for hand use, as shown in the above cut, but can be made to run by power if desired.

Cutting Discs can be applied to this beader so that it can be used as a circular or slitting shear.

No. 0. Beader complete, with one pair Rolls, weighs 650 lbs., ...		**$150 00**
Extra Rolls, ...	per pair,	**16 00**
Cutting Discs, ...	per pair,	**18 00**

PATENT BEADING MACHINES.

Stow's Patent Beading Machines.

WITH DUPLICATE PARTS.

The above cut represents an entirely new Beader. It is most carefully made and will prove an excellent and desirable machine. The rod in front is designed as a "guide rest," on which the cylinder or vessel may rest while forming the bead, and revolves without guidance by the hand.

The machine is made with interchangeable parts, in the same manner as our celebrated Encased Machines.

Nos. 1, 2 and 3 are made with the guide rest, and heavy enough to bead No. 20 iron. Crimping rolls can be furnished to fit these machines.

No. 1. 13 inches, with 3 pairs Rollers and Rotary Stand, weighs 117 lbs., . **$32 25**
No. 2. 10 inches, with 3 pairs Rollers and Rotary Stand, weighs 110 lbs., . **31 25**
Size of Beads for Nos. 1 and 2 are as follows: Ogee Bead, 1 inch; Triple Bead, 1 inch; Single Bead, 3/8 inch.

No. 3. 7 1/2 inches, with 3 pairs Rollers and Rotary Stand, weighs 81 lbs., . **26 25**
Size of Beads for No. 3 are: Ogee Bead, 7/8 inch; Triple Bead, 7/8 inch; Single Bead, 5/16 inch.

Extra Wrought Iron Rollers for Nos. 1 and 2,	per pair,	**3 50**
Extra Wrought Iron Rollers for No. 3,	per pair,	**3 00**
Crimping Rolls for Nos. 1 and 2,	per pair,	**6 00**
Crimping Rolls for No. 3,	per pair,	**5 00**
Extra Stands for Nos 1, 2 and 3,	each,	**1 25**
Beader Cranks or Beader Crank Screws,	each,	**50**

PATENT BEADING MACHINES.

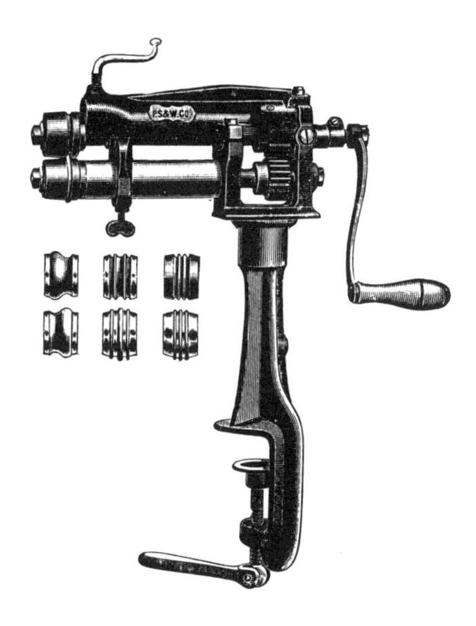

Stow's Patent Beading Machines.

WITH ADJUSTABLE BOXES AND DUPLICATE PARTS.

These Beading Machines are made in the same manner as our Encased Machines. All the parts are made to standard gauges and are lettered, so that any piece can be replaced by designating the number of the Beader and the letter stamped upon the part wanted.

The four pairs of Rollers accompanying the No. 4 Beader are the Single Bead, Ogee Bead, Triple and Triple Coffee Pot, as represented in the above cut.

Crimping Rolls and Guide Rests can be furnished to fit these machines.

No. 4.	6 inches, with four pairs Rollers and Rotary Stand, weighs 35 lbs., . Size of Beads as follows: Ogee, ¾ inch; Triple, ⅝ inch; Triple Coffee Pot, ½ inch; Single, $\frac{3}{16}$ inch.	**$19 75**
No. 5.	4 inches, for Tin, with five pairs Rollers and Rotary Stand, weighs 20 lbs., Size of Beads as follows: Astragal, $\frac{7}{16}$ inch; Ogee, $\frac{7}{16}$ inch; Triple Coffee Pot, ⅜ inch; Triple Coffee Pot, fine, $\frac{5}{16}$ inch; Single Bead, ⅛ inch.	**16 75**

Extra Wrought Iron Rollers for No. 4,	per pair,	**2 00**
Extra Wrought Iron Rollers for No. 5,	per pair,	**1 25**
Crimping Rolls for No. 4,	per pair,	**3 50**
Crimping Rolls for No. 5,	per pair,	**2 00**
Guide Rests for Nos. 4 and 5,	each,	**75**
Extra Stands for Nos, 4 and 5 Beaders,	each,	**75**
Gauge for No. 4 Beader,		**75**
Cap and Crank Screw for No. 4 Beader,		**2 00**

DOUBLE SEAMING MACHINES.

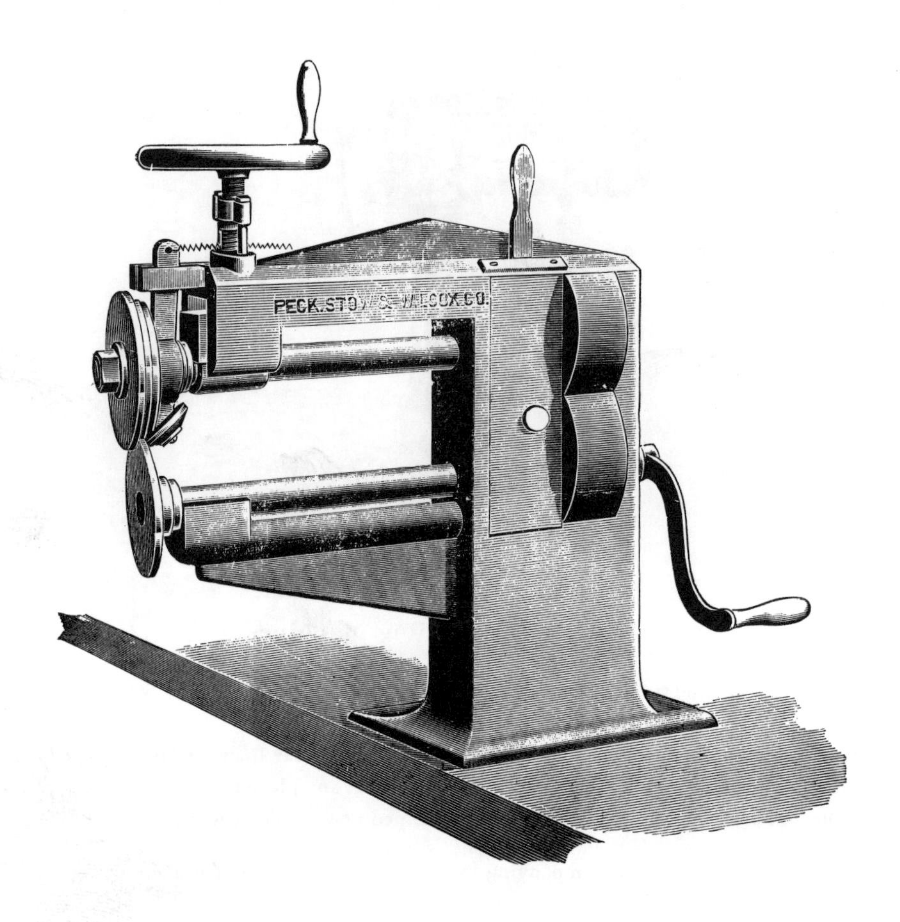

Moore's Patent Double Seaming Machines.

FOR HEAVY METAL.

The No. 00 Moore's Double Seaming Machine will double seam vessels made of metal not heavier than No. 26 Iron, and of a diameter not less than 9¾ inches, and not deeper than 24 inches.

The No. 0 Moore's Double Seaming Machine, illustrated above, will double seam vessels made of metal not heavier than No. 22 Iron, and of a diameter not less than 9¾ inches, and not deeper than 15½ inches.

No.	00. Moore's Double Seamer, weighs 325 lbs., ...	**$175 00**
No.	0. Moore's Double Seamer, weighs 300 lbs., ...	**120 00**

DOUBLE SEAMING MACHINES.

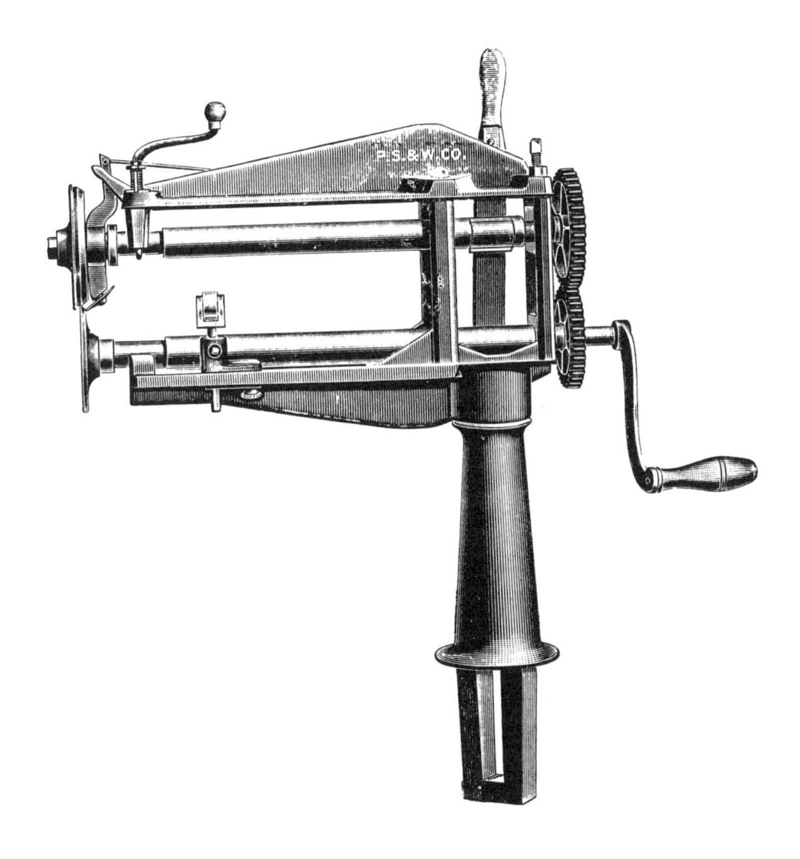

Moore's Patent Double Seaming Machines.

These Double Seaming Machines are adapted for general use. They are well suited to the wants of tinsmiths who desire a machine adapted to different kinds of work. They are the only machines that will double seam coffee pots. The size of the lower face determines the smallest size circle that can be double seamed. Nos. 1 and 2 are with 4-inch lower face and No. 3 has a 3-inch lower face.

No. 1.	15 inches, for heavy metal, 4-inch face, with stand, weighs 50 lbs.,	.	$21 00
No. 2.	13 inches, for common work, 4-inch face, with stand, weighs 45 lbs.,	.	19 00
No. 3.	10 inches, for coffee pots, 3-inch face, with stand, weighs 40 lbs.,	.	16 00
Extra Faces for Nos. 1, 2 and 3,	. . . .	each,	2 50
Extra Stand for Nos. 1, 2 and 3,	. . . .	each,	1 00

DOUBLE SEAMING MACHINES.

Stow's Patent, with Setting Down Attachment.

ELEVEN DISCS.

The Setting Down attached to this machine will set down any work that can be double seamed on the Machine. It works rapidly and accurately, and an inexperienced workman can use it without difficulty. It is also intended to be used on work that does not require to be doubled seamed, and to a large extent will take the place of the Setting Down in the set of Machines. Eleven Discs are furnished with each machine as follows: Flaring Discs, 4⅛, 4⅜, 5¾, 6¼, 8⅝, 10⅞ inch. Straight Discs 5, 5¾, 6, 6½, 8 inch. Discs of extra size or shape can be furnished to order to fit the Machine.

No. 5. Stow's Double Seaming Machine, with Setting Down, weighs 115 lbs.,		**$28 00**
Extra Discs, ...	each,	**1 25**
Extra Heads, ...	each,	**2 50**

DOUBLE SEAMING MACHINES.

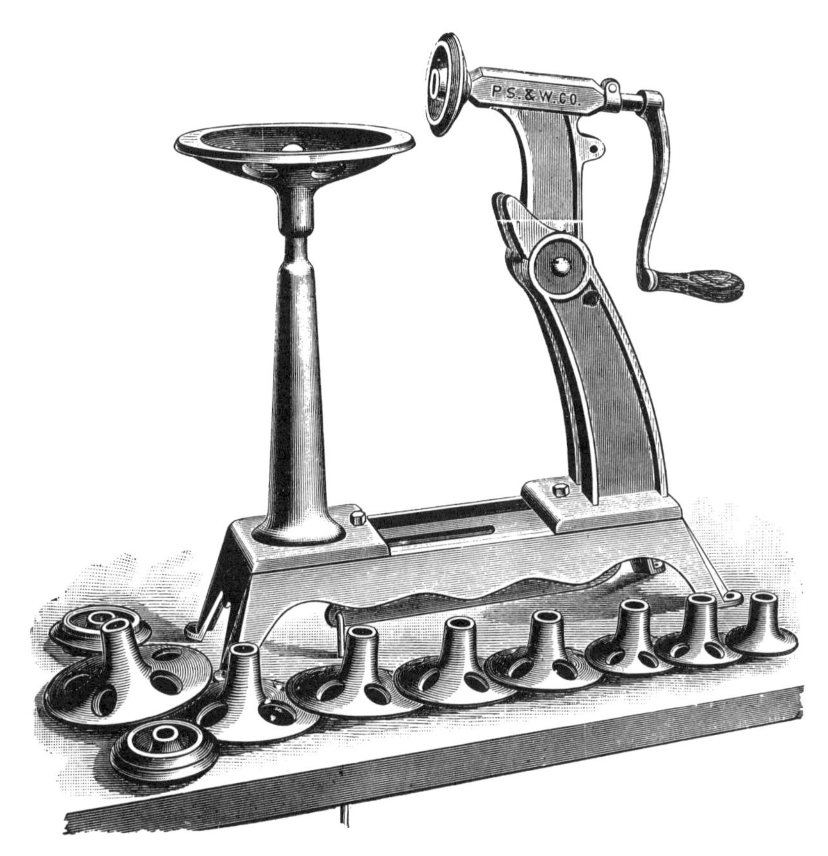

Stow's Patent, Without Setting Down Attachment.

ELEVEN DISCS.

This Machine is more readily adjusted to double seam vessels of different sizes than any other manufactured. It is also the most expeditious in its operation. It acts more favorably in passing seams or locks, because the pressure to set down a double seam is obtained by means of a foot lever, which yields as the roll passes the seam. It is well adapted to Wash Basins, Pans, &c., and will do any work that can be done on similar machines. No. 15 is a machine for which we have had many inquiries, and it is made for heavy work, for which a common double seamer will not answer.

Eleven Discs, as described on the opposite page, are furnished with each machine.

Nos. 15 and 20 can be furnished with setting down attachment if desired.

No. 10.	Double Seaming Machine, without Setting Down, weighs 108 lbs., .	**$24 00**
No. 15.	Double Seaming Machine for heavy work, 20 in. high, weighs 156 lbs.,	**50 00**
No. 20.	Double Seaming Machine for heavy work, 28 in. high, weighs 175 lbs.,	**60 00**
Extra Heads,	. . . . each,	**2 50**

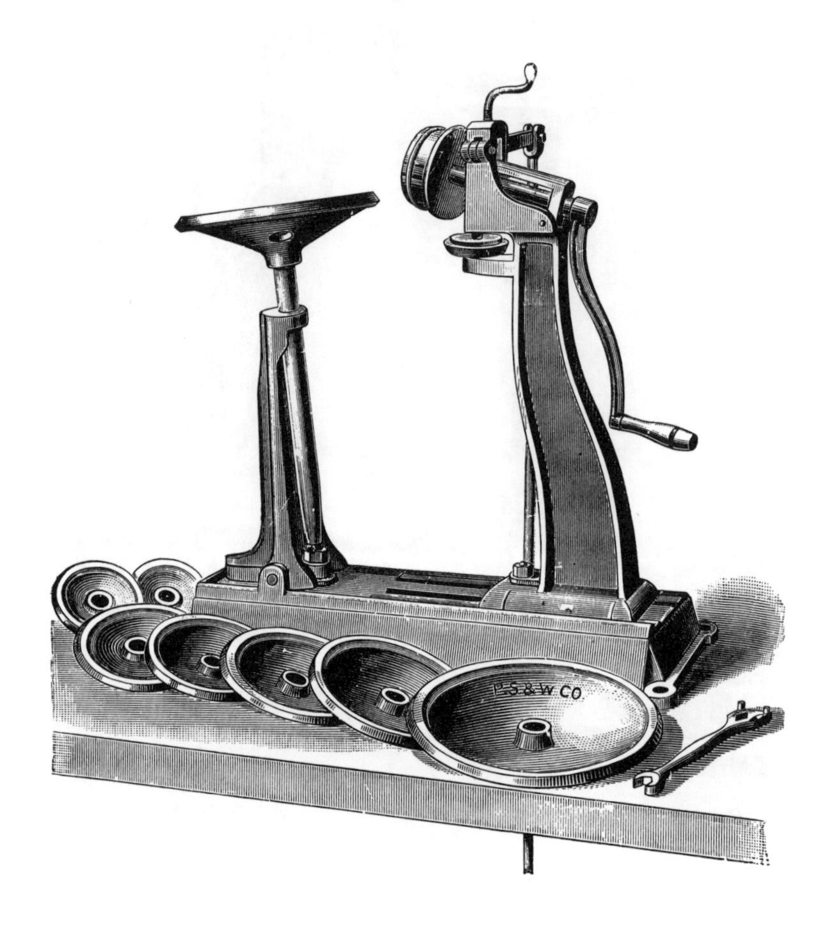

Olmsted's Setting Down and Double Seaming Machine.

EIGHT DISCS.

This machine is 15 inches high; it is adapted for straight or flaring work, and may be used on light or heavy tin. Eight Discs are furnished with each machine, as follows: Flaring Discs, 4½, 6¼, 7¼, 8¾, 11 inch. Straight Discs, 4, 5¾, 8⅛ inch. Olmsted's Double Seamers are now all made to use with treadle or crank screw as represented in cuts.

No. 1. Olmsted's Double Seamer, with Setting Down, weighs 102 lbs., . .		**$30 00**
Extra Discs, ...	each,	**2 00**

DOUBLE SEAMING MACHINES.

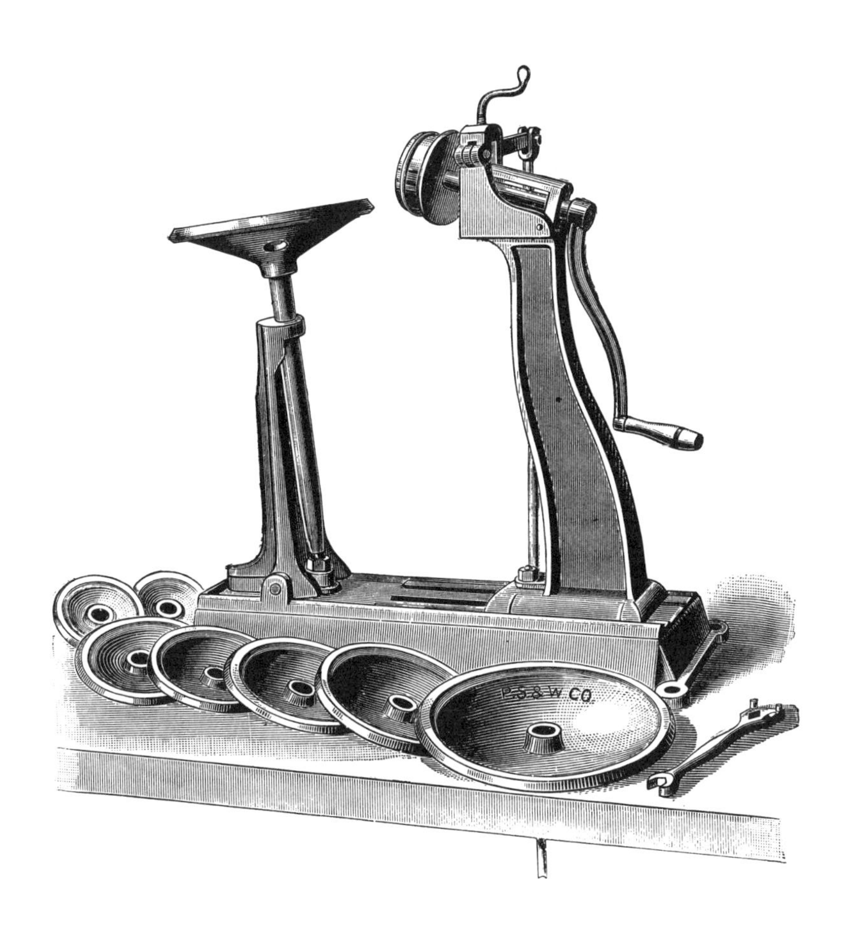

Olmsted's Double Seamer, without Setting Down Attachment.

EIGHT DISCS.

This Machine is similar to the one on the opposite page, but is without the Setting Down Attachment. It is 15 inches high and is adapted to the same kind and variety of work as No. 1. Eight Discs, as described on the opposite page, are furnished with each machine.

No. 2. Olmsted's Double Seamer, without Setting Down, weighs 95 lbs., . .		**$25 00**
Extra Discs, ...	each,	**2 00**

DOUBLE SEAMING MACHINES.

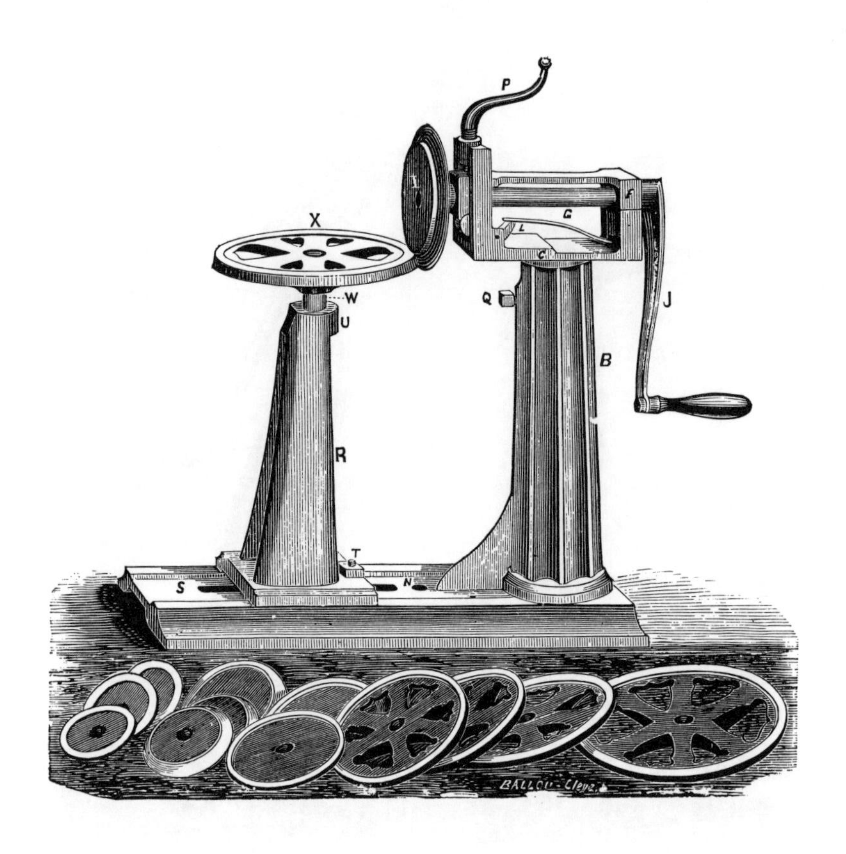

Hulbert's Patent Double Seaming Machine.

TEN DISCS—FOUR FACES.

This machine is adapted to all kinds of flaring and straight work, Coffee and Tea Pots, Oval and Round Boilers, and raised work. It deflects the ware after it is soldered, thereby adding both strength and durability to the article. Ten Discs are furnished with each machine as follows: Flaring Discs, 4⅛, 7⅛, 8⅞, 11½ inch. Straight Discs, 4¼, 5¾, 6¼, 8⅛, 10¼ inch. Oval Edge Disc, 5⅞ inch,

Extra Discs of any size and bevel and duplicate parts of the Machine furnished to order.

No. 1. Hulbert's Double Seamer, with Deflector, weighs 107 lbs.,		**$30 00**
No. 2. Hulbert's Double Seamer, without Deflector, weighs 107 lbs.,		**25 00**
Extra Discs,	each,	**2 00**
Extra Base and Standard,	each,	**8 00**

DOUBLE SEAMING MACHINES.

Burton's Patent Double Seaming Machine.

NINE DISCS.

A Deflector for stiffening Bottoms and Nine Discs for straight or flaring work accompany each Machine. This Machine is especially adapted for Pans and Pails. In ordering discs specify the number.

Discs for flaring work are as follows: No. 1, 11 in.; No. 2, 8⅞ in.; No. 3, 8⅜ in.; No. 5, 7¼ in.; No. 7, 5⅞ in.; No. 9, 4⅛ in.

Discs for straight work are: No. 4, 8⅛ in.; No. 6, 6 in.; No. 8, 5⅝ in.

Special size discs at special prices.

Burton's Double Seamer, complete, weighs 80 lbs., ...		**$24 00**
Extra Discs, ...	each,	**1 25**

WIRE STRAIGHTENERS.

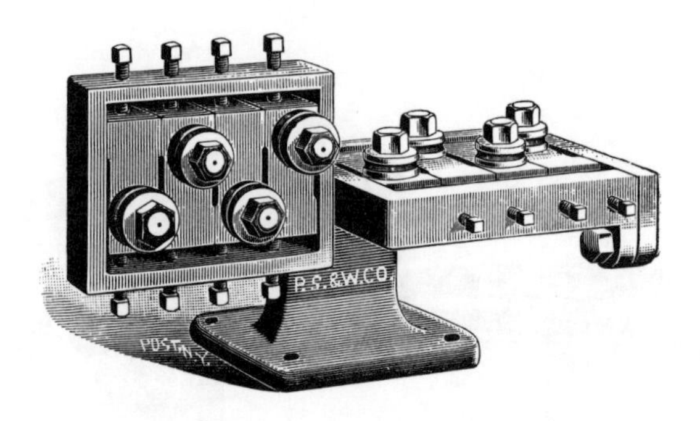

Wire Straightener.

The above engraving shows an Eight Roll Wire Straightener with Central Standard. We can furnish this Straightener with either a horizontal or a vertical stand at the same price.

When this straightener is used alone, the wire is pulled through by hand. We do not recommend it for wire larger than 3/8 inch in diameter.

In ordering a Wire Straightener, please mention the kind and size of wire to be straightened.

No. 5. Eight Roll Wire Straightener, with Central Standard, weighs 45 lbs., . **$40 00**

No. 7. Twelve Roll Wire Straightener, with Central Standard, weighs 52 lbs., . **45 00**

Combined Wire Straightener, Feed and Cutter.

This engraving represents the Wire Straightener, Feed and Cutter combined. When thus made the wire is drawn through the Straightener by means of the Feed, and can be cut to any desired length by means of the Cutter attached.

In the engraving the Machines are fastened together, but they can be used separately by being fastened to the bench in line with each other.

No. 25. Combined Eight Roll Wire Straightener, Feed and Cutter, weighs 86 lbs., **$75 00**

No. 27. Combined Twelve Roll Wire Straightener, Feed and Cutter, weighs 93 lbs., **80 00**

WIRE FEED AND CUTTER.

The Little Gem Wire and Feed Cutter.

The above cut represents a machine that can be used for various purposes. When fastened to the bench it can be used for drawing wire from the coil and cutting it to any desired length.

It can be used as a wire cutter only. By using this Feed and Cutter, wire can be drawn and cut to any desired length with double the rapidity with which it can be cut in any other way.

The size represented is for cutting wire ⅛ inch or less in diameter. Other sizes made to order.

This Machine can be used in connection with Wire Straighteners of our own or other manufacture.

No. 15. Little Gem Wire Feed and Cutter, with Central Standard, weighs 41 lbs., **$40 00**

WIRE REELS.

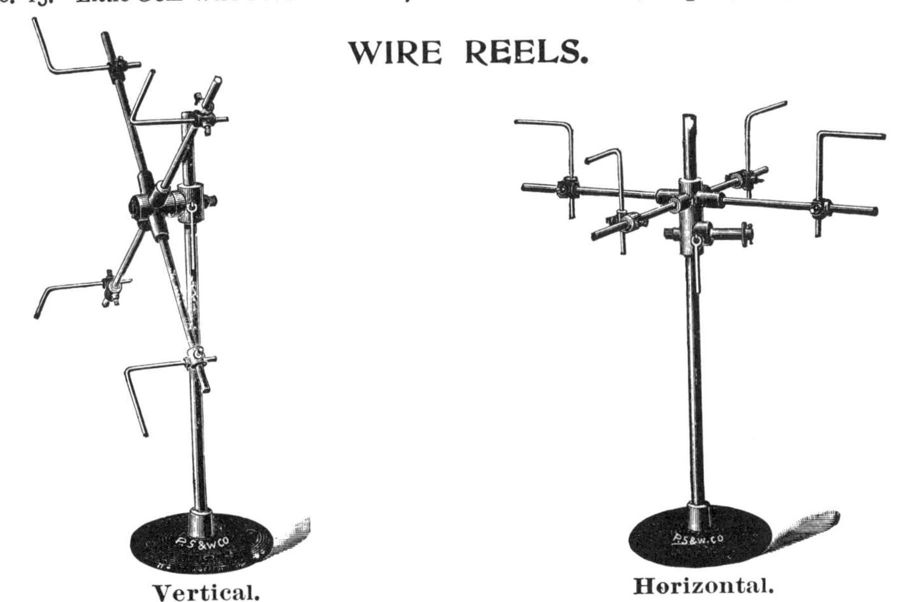

Vertical. **Horizontal.**

The Reel illustrated above is so made that it can be used in a vertical or horizontal position.

The arms can be made of any length desired. As made they are of sufficient length to receive ordinary coils of wire, and can be easily adjusted to receive smaller coils.

No. 11. Wire Reel, weighs 46 lbs., ... **$5 00**

FORMING MACHINES.

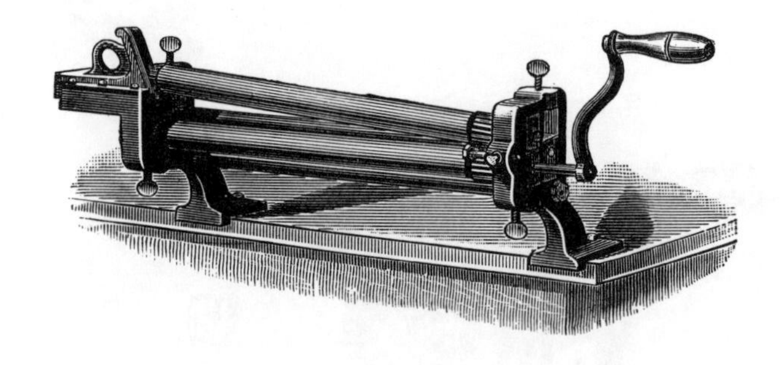

Stow's Improved Tin Pipe Formers.

The Nos. 3 and 4 Tin Pipe Formers are made with slip rolls. They are excellent Machines, with steel rolls and cut gear, and are well adapted for forming speaking tubes and small cylinders.

No. 3.	Tin Pipe Former, 1 inch Steel Rolls, 14 inches long, weighs 37 lbs., .	**$20 00**
No. 4.	Tin Pipe Former, 1 inch Steel Rolls, 20 inches long, weighs 47 lbs., .	**30 00**

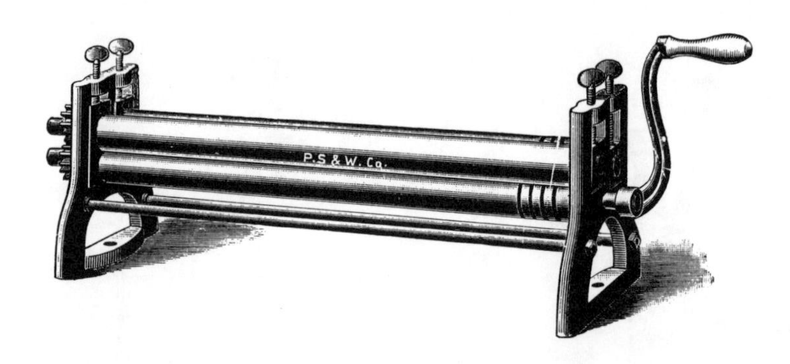

Stow's Patent Stove and Tin Pipe Formers.

These formers are made with steel rolls. They are turned, finely finished, and are free from indentations and imperfections.

No. 0.	For Cans, 2 inch Rolls, 37 inches long, weighs 150 lbs., . .	**$24 00**
No. 0½.	For Cans, 2 inch Rolls, 40 inches long, weighs 155 lbs., ...	**26 00**
No. 1.	Stove Pipe Former, 2 inch Rolls, 30 inches long, weighs 130 lbs., .	**19 00**
No. 2.	Stove Pipe Former, 1¾ inch Rolls, 30 inches long, weighs 105 lbs., .	**18 00**
No. 1.	Tin Pipe Former, 1½ inch Rolls, 20 inches long, weighs 54 lbs., .	**10 00**
No. 2.	Tin Pipe Former, 1½ inch Rolls, 16 inches long, weighs 46 lbs., . .	**9 00**
No. 5.	Tin Pipe Former, 1 inch Rolls, 12 inches long, weighs 34 lbs., .	**12 00**

FORMING MACHINES.

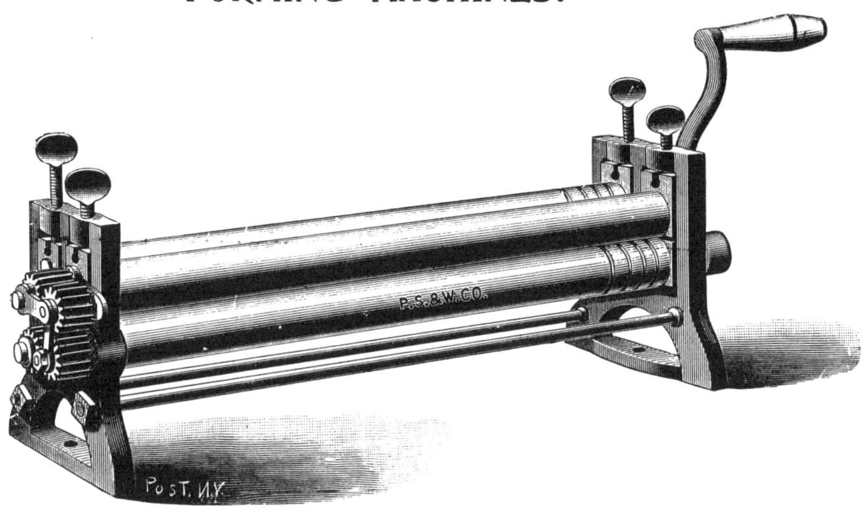

Patent Formers, with Compensating Gear.

The above cut represents our Patent Forming Machines, with compensating gear. Gears, to run smoothly, should mesh constantly to a certain depth, and the shape of the teeth must be accommodated to such depth. In our Patent Formers the cogs attached to the gripping rolls do not articulate with each other, but into the gears, so hung that as the gripping rolls are moved the mesh of the gearing is unchanged, assuring uniformity of action and durability for the cogs; hence, whatever thickness of metal is used the gears run smoothly, without danger of slipping or breaking. The gears are machine cut.

No. 11.	Patent Stove Pipe Former, 1¾ in. Rolls, 30 in. long, weighs 100 lbs.,	**$19 00**
No. 12.	Patent Stove Pipe Former, 2 in. Rolls, 30 in. long, weighs 123 lbs.,	**20 00**
No. 13.	Patent Stove Pipe Former, 2 in. Rolls, 37 in. long, weighs 140 lbs.,	**26 00**
No. 21.	Patent Tin Pipe Former, 1½ in. Rolls, 16 in. long, weighs 42 lbs.,	**10 00**
No. 22.	Patent Tin Pipe Former, 1½ in. Rolls, 20 in. long, weighs 53 lbs.,	**11 00**

Grannis' Patent Slip Roll Formers.

These machines are much more easily adjusted, and can be set for forming any desired size of cylinder in much less time than is required to set others. They are made with Slip Rolls, so that small work can be readily taken off from the ends. On the Nos. 100 and 0100 Formers with 2-inch Rolls, cylinders can be formed from D XXX Tin as small as 2⅛ inches in diameter, and from X Tin as small as 2½ inches. On the No. 200 Former with 1¾ inch Rolls, pipe can be formed from D XXX Tin as small as 2 inches in diameter, and from X Tin 2¼ inches. On the Nos. 01, 300 and 400 Formers, cylinders can be formed from No. 20 iron. They are made with interchangeable parts. Grannis' Slip Roll Formers are now made with patent compensating gear, as described above.

Stove Pipe Formers, complete, with legs to stand on the floor, at an additional cost for each of $7.00.

These Machines can be made to run by power if desired. Pulleys for Nos. 300, 400, 100 or 200, each, 12x4 inches, $8.00.

No. 300.	Former, 2½ in. Rolls, 37 in. long, Back-Geared 2 to 1, weighs 243 lbs.,	**$50 00**
No. 400.	Former, 2½ in. Rolls, 31 in. long, Back-Geared 2 to 1, weighs 225 lbs.,	**45 00**
No. 01.	Former, 2 in. Rolls, 30 in. long, Back-Geared 2 to 1, weighs 154 lbs.,	**35 00**
No. 0100.	Stove Pipe Former, 2 in. Rolls, 37 in. long, weighs 140 lbs., .	**30 00**
No. 100.	Stove Pipe Former, 2 in. Rolls, 30 in. long, weighs 125 lbs., . .	**22 00**
No. 200.	Stove Pipe Former, 1¾ in. Rolls, 30 in. long, weighs 116 lbs., .	**20 00**
No. 101.	Tin Pipe Former, 1½ in. Rolls, 20 in. long, weighs 50 lbs., . .	**12 00**
No. 102.	Tin Pipe Former, 1½ in. Rolls, 16 in. long, weighs 45 lbs., .	**11 00**

FORMING MACHINES.

Stow's Improved Slip Roll Forming Machines.

The Roller around which the work is formed is easily and quickly released from its journal, so that the formed work can be taken from the end of the roller and not sprung over it. This arrangement enables the operator to make the pipe more nearly perfect, and to form even conductor pipe on a Stove Pipe Forming Machine. The Machine is complete in itself, being set on standards or legs.

No. 10. Stove Pipe Former, 2 inch Rolls, 30 inches, weighs 234 lbs., . . . **$26 00**
No. 20. Stove Pipe Former, 1¾ inch Rolls, 30 inches, weighs 210 lbs., . **25 00**

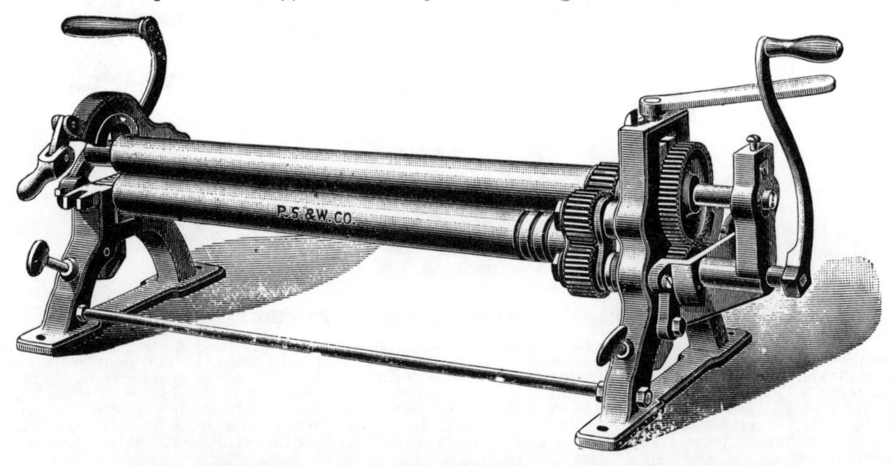

Nos. 000 and 0000 Slip Roll Forming Machines.

These Formers have been especially constructed for forming tubing for Hydraulic Mining or Water Works, where the pressure demands extra heavy iron. They are so made that the roll around which the work is formed is easily raised and held in place, leaving the hands free to slip the formed pipe off the Roll. No. 00 is made in the same manner, except it has not the slip roll. These Formers are now made with compensating gear, as described on page 57, and will form cylinders not less than 12 inches in diameter from iron as stated.

No. 00. Back-Geared 3 to 1, 3 inch Rolls, 37 inches, will form No. 16 Iron, weighs 360 lbs., ... **$75 00**
No. 000. Back-Geared 3 to 1, 3 inch Slip Rolls, 37 inches, will form No. 14 Iron, weighs 450 lbs., ... **90 00**
No. 0000. Back-Geare[illegible] to 1, 4 inch Slip Rolls, 48 inches, will form No. 12 Iron, weighs [illegible] lbs., ... **160 00**

FORMING MACHINES.

Former and Beader.

The above cut represents a machine for forming and shaping plastic material. The beading rolls are movable and can be used in different positions on the rolls.

No. 67. With Steel Slip Rolls, 20 inches long, ... **$26 00**

Extra Heavy Formers on Legs.

The above cut represents either our No. 000 or 0000 Former on Legs, with Pulley for power. They are adapted for the same work and the same thickness of metal as described on the opposite page.

No. 000. With Pulley, 3-inch Steel Slip Rolls, 37 inches long, Back-Geared 3 to 1, for Power and on Legs, weighs 600 lbs., ... **$110 00**

No. 0000. With Pulley, 4-inch Steel Slip Rolls, 48 inches long, Back-Geared 4 to 1, for Power and on Legs, weighs 1,000 lbs., . . . **180 00**

FORMING MACHINES.

Forming and Beading Machine.

The above cut represents a Machine for forming and beading strips or narrow widths of metal as wide as 6 inches, such as milk can bands, furnace bands, etc. Will form cylinders as small as 3 inches in diameter.

We make a great variety of Machines similar to those here represented, and will be pleased to quote prices for special work required.

No. 62. With Steel Slip Rolls, 6 inches long, . . , . . **$30 00**

Extra Heavy Former.

The above cut represents an Extra Heavy Former, for special work. It is adapted to forming cylinders from 4 inches in diameter, upward, from stock as thick as ⅛ inch, and it will be found very desirable for such work.

We can make them in different lengths to suit the wants of our customers.

No. 52. 3-inch Steel Slip Rolls, 12 inches long, Back-Geared 10 to 1, . **$125 00**

FORMING MACHINES.

The following Patent Slip-Roll Forming Machines are made with steel rolls and cut gear. They are back-geared and will form cylinders not less than 12 inches in diameter from No. 12 Iron or lighter, according to the length and diameter of the rolls.

These machines are well made and well fitted; the rolls are of a superior quality of steel, finely finished, with surfaces free from indentations and imperfections.

The device for releasing and raising the upper roll is quick acting, and of great advantage to the operator in forming cylinders of small diameters.

In addition to the sizes given below, we manufacture a large line of special formers, of different lengths and diameters, and of special designs for special work. Careful estimates will be given when specifications are submitted.

3-inch Steel Slip Rolls, Back-Geared 3 to 1.

Length.	Weight.	List.	Extra for Legs.	Extra for Power.
30 inches.	464 pounds.	$80 00	$8 00	$12 00
36 inches.	500 pounds.	90 00	8 00	12 00
42 inches.	536 pounds.	95 00	8 00	12 00
48 inches.	572 pounds.	100 00	8 00	12 00
54 inches.	608 pounds.	105 00	8 00	12 00
60 inches.	644 pounds.	110 00	8 00	12 00
66 inches.	680 pounds.	115 00	8 00	15 00
72 inches.	716 pounds.	120 00	8 00	15 00

4-inch Steel Slip Rolls, Back-Geared 4 to 1.

Length.	Weight.	List.	Extra for Legs.	Extra for Power.
42 inches.	690 pounds.	$150 00	$12 00	$15 00
48 inches.	830 pounds.	155 00	12 00	15 00
54 inches.	900 pounds.	160 00	12 00	15 00
60 inches.	1000 pounds.	165 00	12 00	15 00
66 inches.	1100 pounds.	170 00	12 00	15 00
72 inches.	1200 pounds.	175 00	12 00	18 00
78 inches.	1325 pounds.	185 00	12 00	18 00
84 inches.	1450 pounds.	195 00	12 00	18 00

CORRUGATING MACHINES.

Corrugating and Crimping Machines.

The above cut represents a Power Crimping Machine on legs, with 5-inch rolls, 24 inches long. It is back-geared 8 to 1. We make these Machines in several sizes. They have cut gear, cast rolls with corrugations planed therein. These Machines are of the highest grade and best quality. We are confident they are unequalled by any similar Machines in the market. They will crimp No. 20 Iron. The crimping can be of any desired space and the corrugations of any depth or style. These Machines can be made to work by hand if desired, or so that either crank or pulley can be used. If made for hand only the list price will be $8.00 less on each machine.

No. 501.	5-inch Rolls, 24 inches long,	$225 00
No. 503.	5-inch Rolls, 30 inches long,	240 00
No. 505.	5-inch Rolls, 36 inches long,	255 00
No. 507.	5-inch Rolls, 42 inches long,	275 00
No. 601.	6-inch Rolls, 24 inches long,	275 00
No. 603.	6-inch Rolls, 30 inches long,	290 00
No. 605.	6-inch Rolls, 36 inches long,	305 00
No. 607.	6-inch Rolls, 42 inches long,	325 00
No. 609.	6-inch Rolls, 48 inches long,	350 00

CORRUGATING MACHINES.

Corrugating and Crimping Machines.

The above cut represents an extra heavy Power Corrugating Machine on legs, with 8-inch cast rolls, 48 inches long. It has cut-gear and will crimp No. 20 iron. They are made in the same manner as those on page 62, that is, they can be arranged so that either crank or pulley can be used. If made for hand only the list price will be $8.00 less on each machine.

No. 701.	7-inch Rolls, 24 inches long,	$350 00
No. 703.	7-inch Rolls, 30 inches long,	375 00
No. 705.	7-inch Rolls, 36 inches long,	400 00
No. 707.	7-inch Rolls, 42 inches long,	425 00
No. 709.	7-inch Rolls, 48 inches long,	450 00
No. 711.	7-inch Rolls, 54 inches long,	475 00
No. 713.	7-inch Rolls, 60 inches long,	500 00
No. 801.	8-inch Rolls, 24 inches long,	425 00
No. 803.	8-inch Rolls, 30 inches long,	450 00
No. 805.	8-inch Rolls, 36 inches long,	475 00
No. 807.	8-inch Rolls, 42 inches long,	500 00
No. 809.	8-inch Rolls, 48 inches long,	525 00
No. 811.	8-inch Rolls, 54 inches long,	550 00
No. 813.	8-inch Rolls, 60 inches long,	575 00
No. 815.	8-inch Rolls, 66 inches long,	600 00
No. 817.	8-inch Rolls, 72 inches long,	650 00

CORRUGATING MACHINES.

Corrugating and Crimping Machines.

The above cut represents a Power Corrugating or Crimping Machine, with 3-inch steel rolls and cut gear, 24 inches long. It was designed for special work and gives the best satisfaction. It is back-geared 8 to 1. Will corrugate IXXX Tin.

We can furnish other styles and sizes. The crimping can be of any desired space and the corrugation of any depth These machines can be made to work by hand if desired, or so that either crank or pulley can be used. If made for hand only the list price will be $8.00 less on each machine.

No. 301.	3-inch Steel Rolls, 8 inches long, weighs 143 lbs., ...	**$100 00**
No. 303.	3-inch Steel Rolls, 12 inches long, weighs 176 lbs., . -	**110 00**
No. 305.	3-inch Steel Rolls, 15 inches long, weighs 200 lbs., . . .	**115 00**
No. 307.	4-inch Steel Rolls, 20 inches long, weighs 520 lbs., . .	**125 00**
No. 309.	4-inch Steel Rolls, 24 inches long, weighs 550 lbs., . . .	**130 00**

FORMING MACHINES.

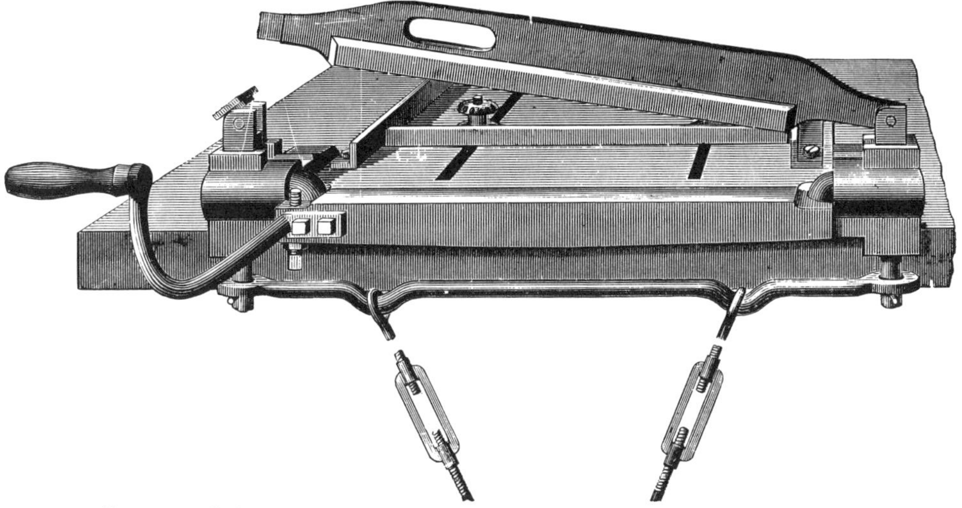

Improved Square Box and Square Pipe Forming Machines.

No. 00 will form work 36 inches long, 3 inches square and larger. No. 0 will form work 30 inches long, 3 inches square and larger. No. 1 will form work 20 inches long, 2½ inches square and larger, and will form at any angle greater than a right angle. No. 2 will do the same work as No. 1, but is only 15 inches long.

The Clamping Bar is so arranged that it can be lifted and the formed work can be slipped off the end.

No. 00.	Square Box Machine, 36 inches long, weighs 225 lbs.,	$75 00
No. 0.	Square Box Machine, 30 inches long, weighs 190 lbs.,	50 00
No. 1.	Square Box Machine, 20 inches long, weighs 111 lbs.,	30 00
No. 2.	Square Box Machine, 15 inches long, weighs 90 lbs.,	20 00

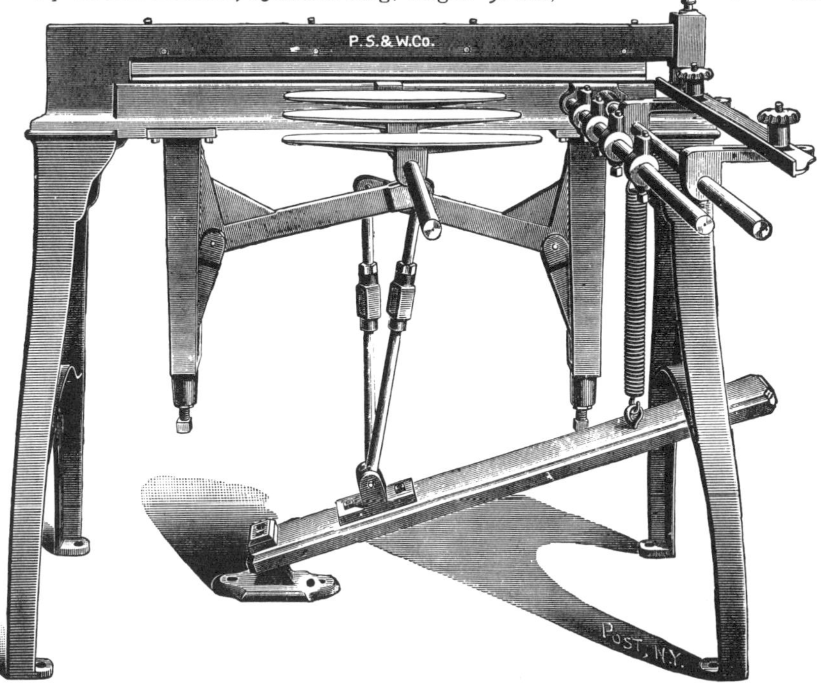

Improved Angular Box and Pipe Forming Machine.

This machine is operated by foot and works easily for making triangular, square, hexagonal, or any other shaped vessel, having three or more sides that are regular or irregular, corners of which can be left sharp or circular as desired. If tubes are to be formed for heating purposes, they can be made tapering, so as to go easily together. It will form a square pipe or box from light sheet metal three inches square or larger, 30 in. long. Tapering pipe can also be formed, the small end of which is of a diameter not less than 3 in.

No. 10. Angular Box Machine, 30 inches long, on Legs, weighs 225 lbs., . $40 00

TUBE FORMERS.

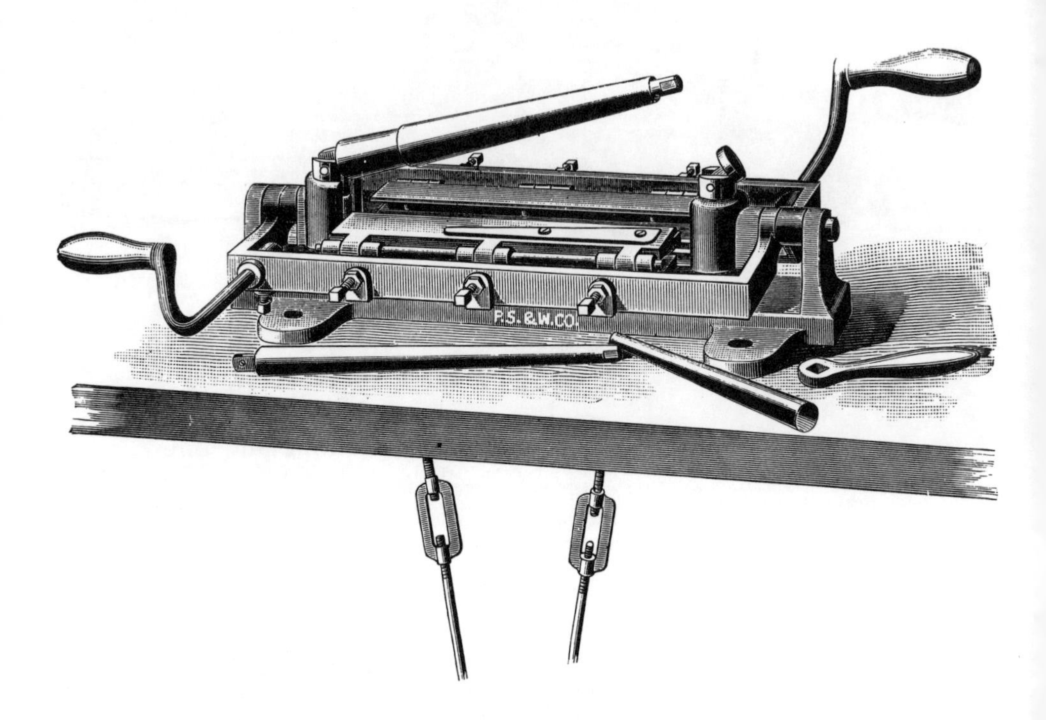

Stow's Patent Tube Formers.

These Tube Formers are well made and accurately fitted and will form tubes of the following diameters: Nos. 000, 00 and 0, ⅝ to 2 inches; Nos. 1 and 2, ½ to 2 inches; No. 3, ⅜ to 1¾ inches. With extra rods and extra wings the same Tube Former can be made to form different size Tubes.

Where no size of rod is specified in the order, 1 inch rod will be sent with each No. 000, 00, 0 and 1.

A large variety of rods and beds for special work can be furnished. Prices given on receipt of sample or description of work.

No. 000.	To form Speaking Tubes 24 inches long, weighs 100 lbs.,	$60 00
No. 00.	To form Tubes 20 inches long, weighs 80 lbs.,	45 00
No. 0.	To form Tubes 15 inches long, weighs 74 lbs.,	30 00
No. 1.	For Candle Moulds or Ladle Handles, 11 inches long, weighs 50 lbs.,	25 00
No. 2.	For Tea Kettle Spouts, etc., 8 inches long, weighs 43 lbs.,	20 00
No. 3.	For Rattle Box Handles, 5 inches long, weighs 37 lbs.,	20 00
Extra Die Rod for No. 000,		5 00
Extra Die Rod for No. 00,		4 00
Extra Die Rod for No. 0,		3 00
Extra Die Rod for No. 1,		2 00
Extra Die Rod for Nos. 2 and 3,	each,	1 50

TUBE FORMERS.

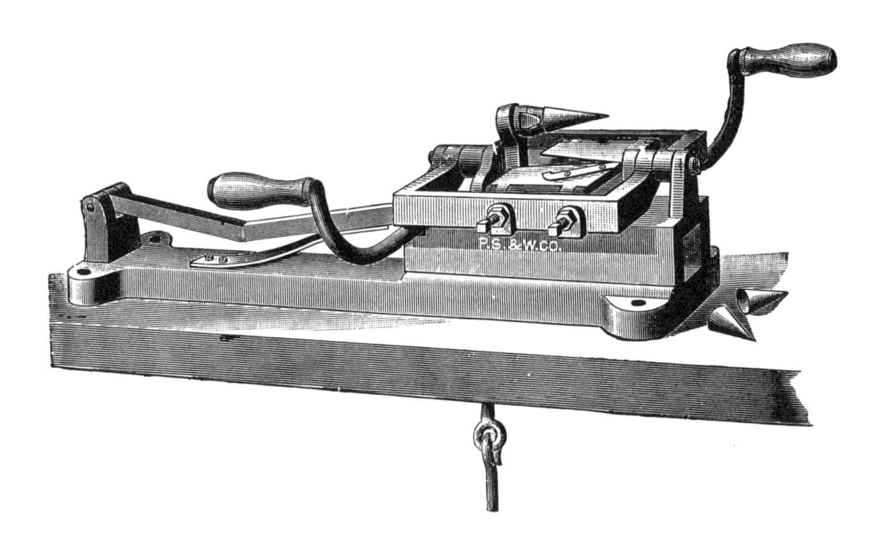

Stow's Patent Tube Formers.

This Tube Former is constructed for forming straight or tapering tubes. We make a large variety for special work.

We manufacture a Patent Tube Former that works by power, that will form tubes very rapidly from light material, as small as ¼ inch in diameter.

No. 4. For Tubes 4¾ inches long, not over 2 inches in diameter, weighs 50 lbs.,	**$25 00**
Extra Die Rod and Bed for No. 4, . . , . . , .	**1 50**

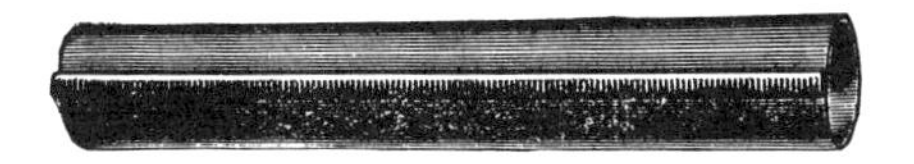

Knurled Tube.

We manufacture a set of Machines constituting A COMPLETE OUTFIT for making straight Tubes with lock-seams from 1 to 10 feet long and from ⅞ to 3 inches in diameter. Will be pleased to quote prices on receipt of sample indicating shape, style and description of work.

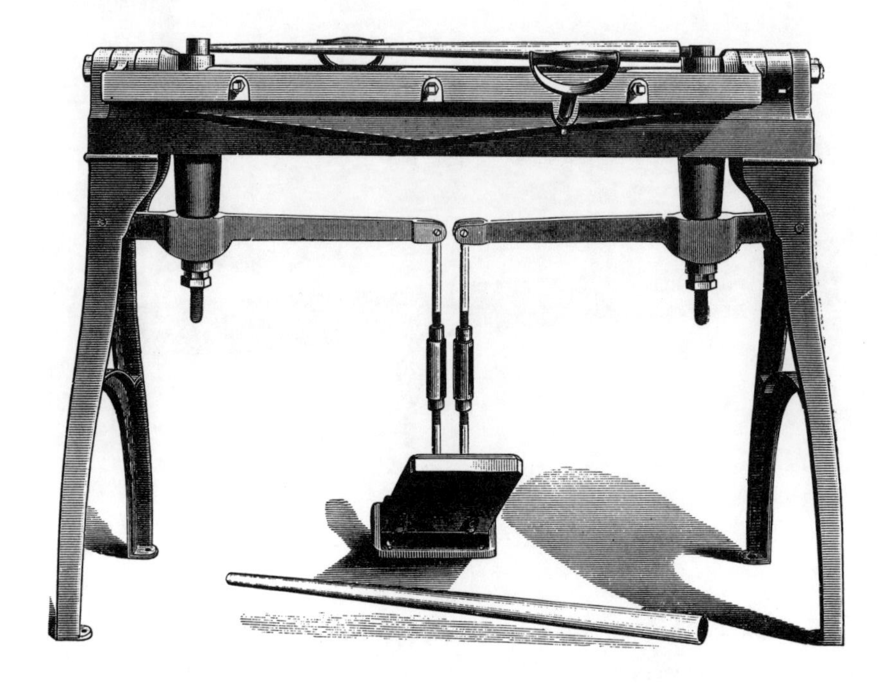

Stow's Patent Tube Former, on Legs.

The above cut represents our No. 0000 Tube Forming Machine. It embraces all the latest devices and improvements which we have introduced in this class of machines.

This Machine is constructed to form either straight or tapering tubes 32 inches long; but will not form straight tubes to advantage of a diameter less than 3/4 inch.

No. 0000. Tube Former, to form 3/4 to 3 in. diameter, 32 in. long, weighs 440 lbs., **$125 00**

TUBE FORMERS.

Spaulding's Patent Tube Formers.

The Machine represented in the above cut can be used only with power, but can be arranged to work by hand. It is so constructed that the same Machine will form tubes or cylinders 24 inches long and of diameters ¾ to 1¾ inches, and any intermediate size. This Machine is specially adapted for forming speaking tubes.

We make to order a large variety of special Tube Formers for different kinds of work, and can furnish Formers to make short tubes $\frac{3}{16}$ of an inch in diameter. Tubes 10 inches in length can be made as small as $\frac{5}{16}$ of an inch in diameter.

When ordering it is desirable to send a sample of the tube wanted.

No. 13.	Spaulding's Tube Former, with one Forming Roll, weighs 362 lbs.,	$75 00
No. 14.	Spaulding's Tube Former, with one Forming Roll, for hand power, weighs 350 lbs., ...	70 00
No. 15.	Spaulding's Tube Former, with one Forming Roll, extra size, to form large Fish Horns, weighs 400 lbs., ...	125 00
Extra Forming Rolls for No. 13,	. . . . , . . .	1 75

FORMING MACHINES.

Wire Ring Former.

Wire Ring Former or Winder, weighs 14 lbs., ... **$3 00**

Candlestick Former and Beader, with Standard.

Blacking, Pepper or Rattle Box and Candlestick Former and Beader, with Steel Rolls and Stand, weighs 30 lbs., ... **$20 00**

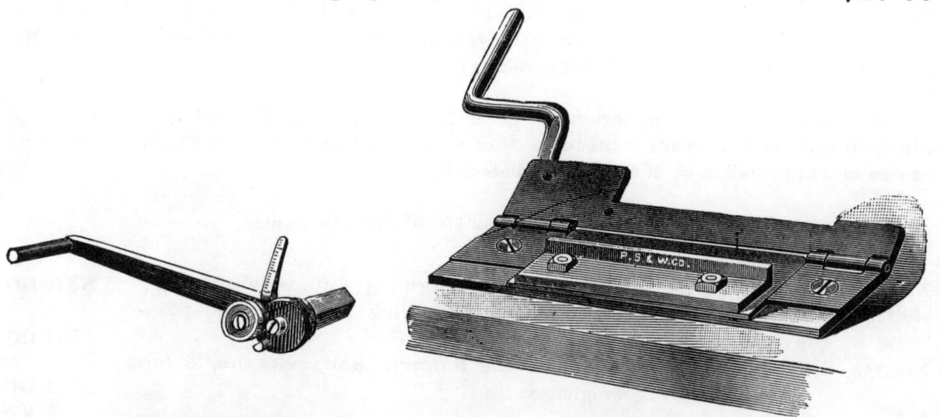

Wire Bail Former. **Cleat Folder.**

Miller's Wire Bail Former, for forming Wire Bails for Pails, weighs 2 lbs., . **$1 50**

Miller's Cleat Folder, for making Cleats for Roofing, weighs 2 lbs., . . **2 00**

FORMING MACHINES.

Miller's Patent Oval Handle Former.

An adjustable Gauge enables the operator to form different size Handles.

Miller's Oval Handle Former, for forming Oval Rings of Wire, weighs, 3 lbs., . **$3 00**

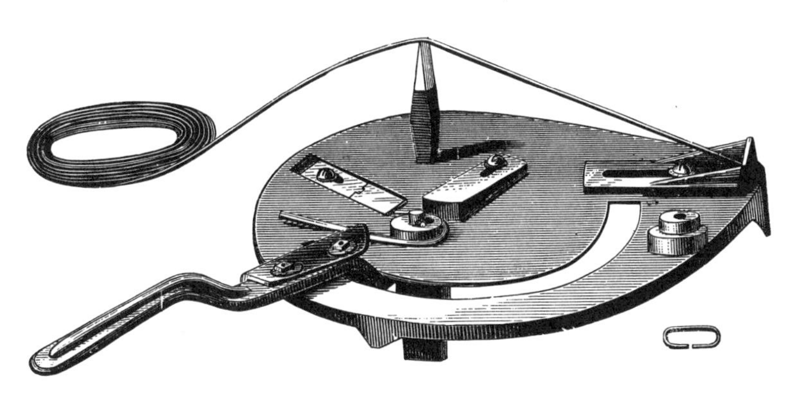

Jones' Patent Oval Handle Former.

A Gauge to measure the wire and three sets of Rolls for forming different size handles are attached to each Machine.

Jones' Oval Handle Former, for forming Oval Rings of Wire, weighs 9 lbs., . **$6 00**

Smith's Patent Oval Handle Former.

Three sets of Rolls for forming different size Handles accompany each Machine.

Smith's Oval Handle Former, for forming Oval Rings of Wire, weighs 9 lbs., . **$6 00**

Boss Former.

Boss Former, for forming Handle Braces, weighs 6 lbs., ... **$1 50**

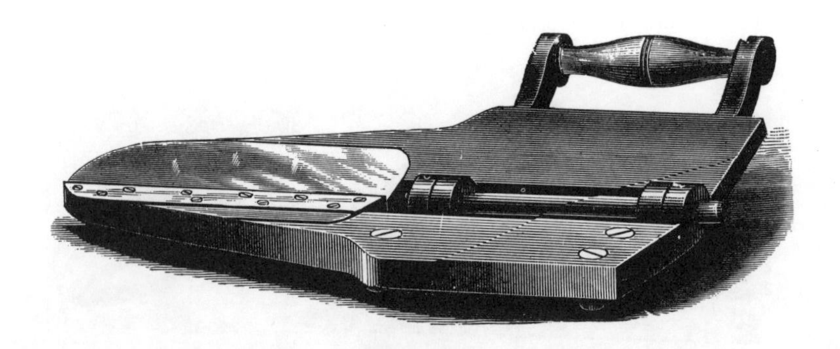

Can Top Folding Machine.

Can Top Folding Machine, for Folding Locks for Tops of Oil Cans, weighs 23 lbs., **$8 00**

FORMING MACHINES.

Van Bramer's Patent Wire Cutter and Bail Former.

A simple, cheap, durable and economical Machine. It takes wire from the coil, and gauges and cuts it to the desired length. It cuts smoothly and easily *one quarter inch wire*, as well as all smaller sizes. It forms Bucket Bails with rapidity and accuracy. It saves the vexations experienced by every Tinsmith in the frequent breaking of Cutting Nippers.

Van Bramer's Wire Cutter and Bail Former, weighs 30 lbs., ... **$10 00**

BOILER EXPANDER.

Patent Boiler Expander.

The above cut represents a new and useful invention, designed to hold the bodies of wash-boilers in their place while the bottoms are being double seamed.

No. 5. Boiler Expander, weighs 2 lbs., ... **$1 75**

SHEET IRON FOLDING MACHINES.

Wright's Patent Sheet Iron Folders.

WILL TURN LOCKS $\frac{3}{16}$ AND $\frac{5}{16}$ INCH.

No. 00. Wright's Patent Sheet Iron Folder, 62 inches, weighs 133 lbs., . **$50 00**
No. 0. Wright's Patent Sheet Iron Folder, 42 inches, weighs 96 lbs., . **20 00**
No. 1. Wright's Patent Sheet Iron Folder, 30 inches, weighs 64 lbs., . **12 00**

As now made this Machine will turn locks of different widths on any length of sheet. To turn an edge longer than the length of the Machine, use the round rod, placing it so that the sheet will be put into the Machine over it and turn the folding bar against its round surface, making a slight bend the entire length of the sheet, and repeat this operation of bending until the lock is finished, or far enough to close it down in the ordinary way. In turning edges on long sheets it will be well to turn the edge slightly at first and repeat the process of bending until finished. To turn wider locks than the depth of the folding plate use the steel strips under the plate to increase the width, and operate the Machine in the same manner as without them.

Wright's Improved Sheet Iron Folders.

WILL TURN LOCKS $\frac{3}{16}$, $\frac{5}{16}$, ½ AND ⅝ INCH.

No. 12. Wright's Improved Sheet Iron Folder, 30 inches, weighs 69 lbs., . **$15 00**
No. 17. Wright's Improved Sheet Iron Folder, 62 inches, weighs 150 lbs., . **50 00**

SHEET IRON FOLDING MACHINES.

Stow's Improved Sheet Iron Folders.

WILL TURN LOCKS $\frac{3}{16}$ TO $\frac{3}{8}$ INCH.

These Machines are so constructed that the Gauge always moves upon a line parallel with the edge of the Folding Plate. The principle of folding is the same as in the Wood Bottom Folder.

No. 10.	Stow's Sheet Iron Folder, 30 inches, weighs 58 lbs., . .	**$10 00**
No. 15.	Stow's Sheet Iron Folder, 42 inches, weighs 175 lbs., . .	**18 00**
No. 20.	Stow's Sheet Iron Folder, 60 inches, weighs 300 lbs., . .	**25 00**

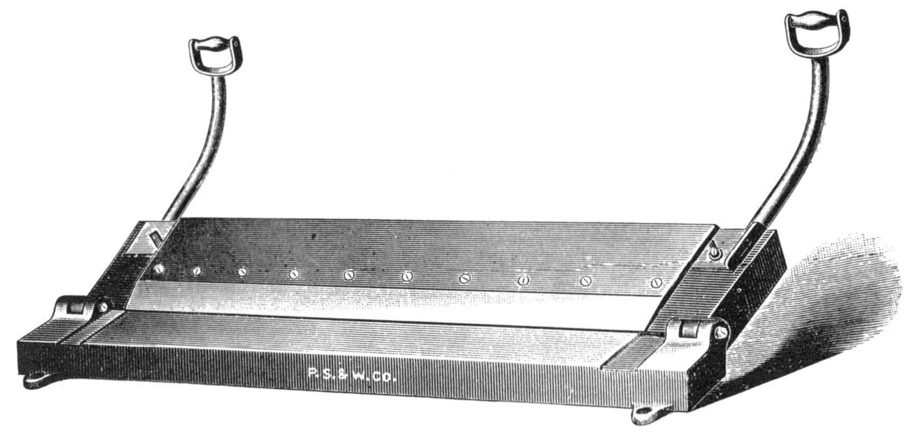

Stow's Improved Sheet Iron Folders.

WILL TURN LOCKS $\frac{1}{4}$ TO $2\frac{1}{4}$ INCHES.

These Machines are constructed on the same principle as those named above, but are extra heavy and will turn edges on No. 20 iron, from $\frac{1}{4}$ to $2\frac{1}{4}$ inches in width, and at any angle. They can be constructed to turn wider edges if so ordered.

No. 25.	Stow's Sheet Iron Folder, 42 inches, weighs 290 lbs., . .	**$80 00**
No. 30.	Stow's Sheet Iron Folder, 48 inches, weighs 325 lbs., ...	**90 00**
No. 35.	Stow's Sheet Iron Folder, 60 inches, weighs 400 lbs , . .	**125 00**

SHEET IRON FOLDING MACHINES.

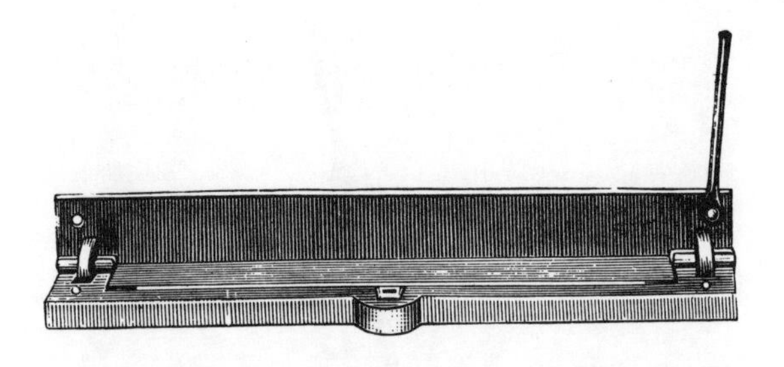

Iron Bottom Sheet Iron Folders.

WILL TURN LOCKS $\frac{3}{16}$ TO ½ INCH.

No. 2.	Iron Bottom Sheet Iron Folder, 30 inches, weighs 53 lbs., . .	**$7 00**
No. 3.	Iron Bottom Sheet Iron Folder, 39 inches, weighs 68 lbs., .	**12 00**

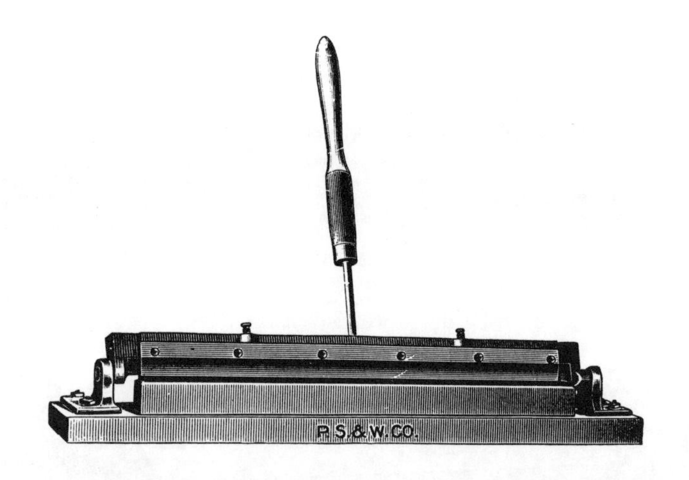

Wood Bottom Sheet Iron Folders.

WILL TURN LOCKS $\frac{3}{16}$ TO ½ INCH.

We are now constructing this folder with a milled bar and are fitting **the gauge accurately,** so that it can be used for Tin as well as Sheet Iron.

No. 0.	Wood Bottom Sheet Iron Folder, 48 inches, weighs 125 lbs., .	**$20 00**
No. 1.	Wood Bottom Sheet Iron Folder, 30 inches, weighs 56 lbs., , .	**10 00**

GUTTER MACHINES.

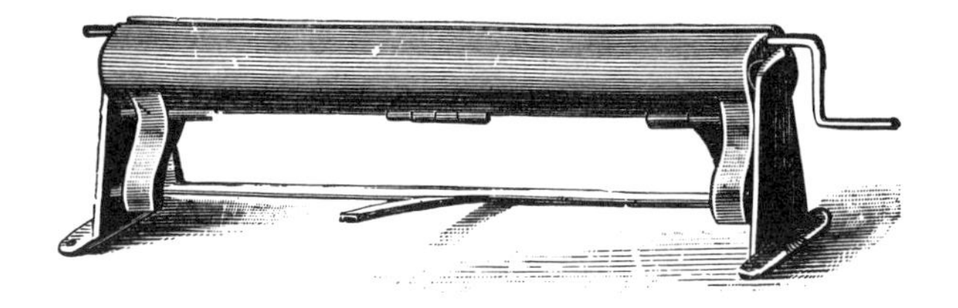

Stow's Patent Gutter Beaders.

All Gutter Beaders longer than 30 inches have two handles, so that the rod can be turned from both ends to prevent it from twisting. Special lengths of Stow's Patent Gutter Beaders can be furnished to order.

No. 00.	Steel Rod, 72 inches long, ⅞ or 1 inch diameter, weighs 100 lbs.,	$35 00
No. 0.	Steel Rod, 30 inches long, ½ to ¾ inch diameter weighs 66 lbs.,	10 00
No. 1.	Steel Rod, 20 inches long, ⅜ to ⅝ inch diameter, weighs 30 lbs.,	5 50
No. 2.	Steel Rod, 15 inches long, $\frac{5}{16}$ to ½ inch diameter, weighs 17 lbs.,	4 50

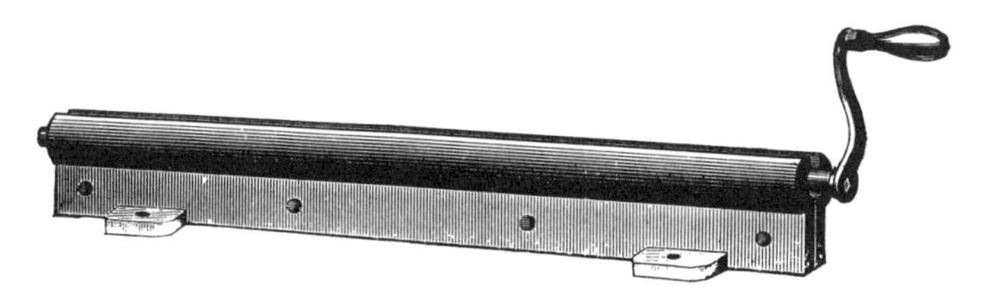

Iron Bottom Gutter Beaders with Enclosed Rod.

No. A3.	Iron Bottom, with Enclosed Rod, 120 inches, weighs 200 lbs., .	$40 00
No. A2.	Iron Bottom, with Enclosed Rod, 96 inches, weighs 160 lbs., . .	30 00
No. A1.	Iron Bottom, with Enclosed Rod, 60 inches, weighs 90 lbs., .	20 00
No. 01.	Iron Bottom, with Enclosed Rod, 42 inches, weighs 25 lbs., . .	9 00
No. 02.	Iron Bottom, with Enclosed Rod, 30 inches, weighs 18 lbs., . .	6 00
No. 11.	Iron Bottom, with Enclosed Rod, 20 inches, weighs 11 lbs., . .	4 00
No. 12.	Iron Bottom, with Enclosed Rod, 15 inches, weighs 6 lbs., . .	3 50

In ordering the above Gutter Beaders be sure and specify the diameter of Rod wanted.

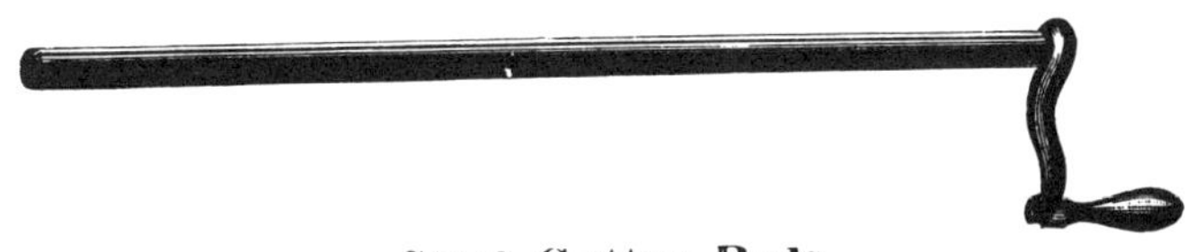

Steel Gutter Rods.

No. A3.	Steel Gutter Rods, ¾ to 1 inch diameter, 120 inches long, . .	$20 00
No. A2.	Steel Gutter Rods, ⅝ to 1 inch diameter, 96 inches long, . .	16 00
No. A1.	Steel Gutter Rods, ⅝ to 1 inch diameter, 60 inches long, . .	8 00
No. 01.	Steel Gutter Rods, ½ to ⅞ inch diameter, 42 inches long, . .	4 50
No. 02.	Steel Gutter Rods, ½ to ¾ inch diameter, 30 inches long, . .	3 00
No. 11.	Steel Gutter Rods, ⅜ to ⅝ inch diameter, 20 inches long, . .	2 00
No. 12.	Steel Gutter Rods, ⅜ to ⅝ inch diameter, 15 inches long, . .	1 75

Steel Gutter Rods of other diameters and lengths made to order.

SQUARING SHEARS.

Stow's Power Squaring Shears.

The Power Shears herewith illustrated are of great capacity. The gate or cutter-bar is very strong and slides on well fitted ways, which are provided with gibs for adjustment to compensate for wear. An automatic clutch is fastened to the main shaft which allows it to make but one revolution, always stopping the cutter bar at its highest point. In front of the cutting blades is a clamping bar, made adjustable for different thicknesses of metal, which holds the sheet in a firm position while being cut. The gears are machine cut and have very broad faces. With this machine are our patented back extension arms with verneir adjustment, simple and effective. There are also front extension arms, and the usual back, front and side gauges. It also has an extra back gauge for use in cutting narrow strips. On the top of the bed plate deep lines are cut two inches apart, parallel with the cutting edge of the blades, which facilitates the accurate placing of the bed gauges. We guarantee these shears to cut iron or steel as thick as No. 10 gauge their whole length.

No. 100XX. Will cut No. 10 Iron, 100 inches in length, weighs 6000 lbs., ... $800 00
No. 132XX. Will cut No. 10 Iron, 132 inches in length, weighs 8000 lbs., ... 1200 00

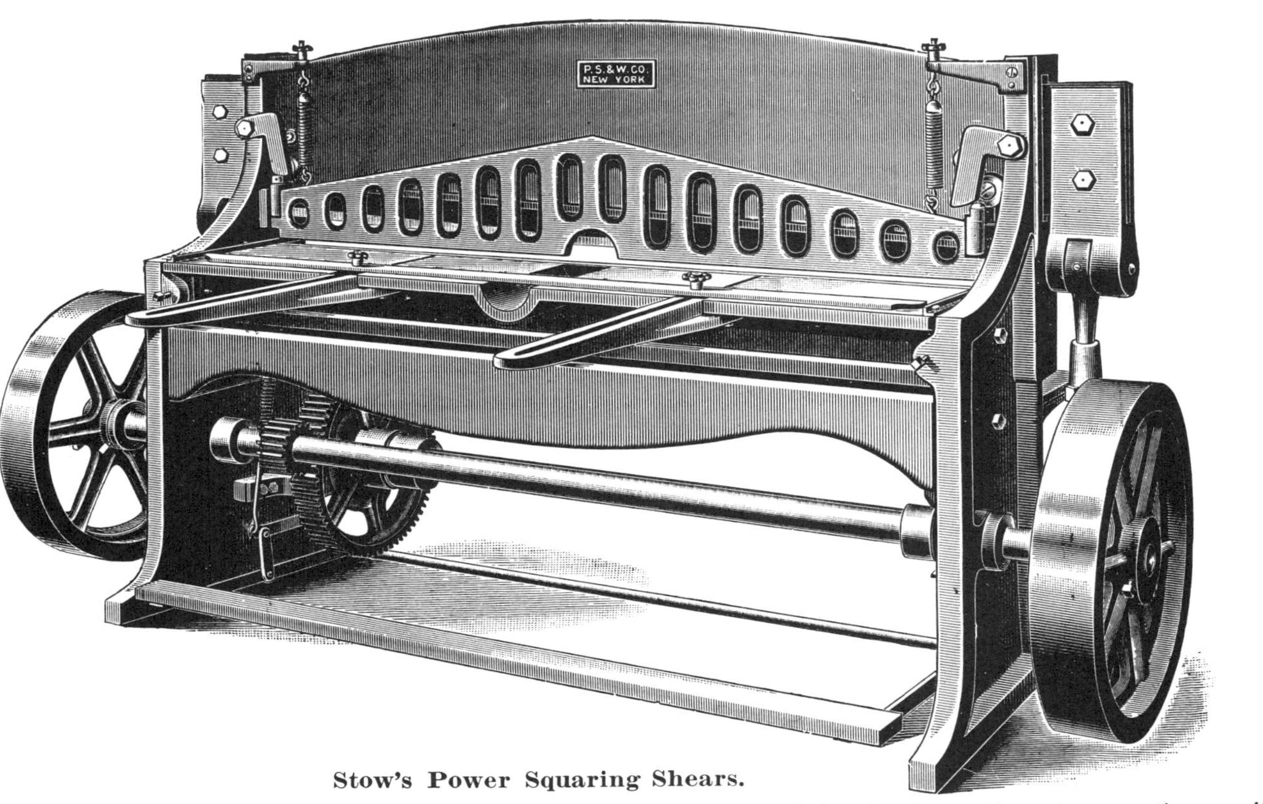

Stow's Power Squaring Shears.

These Shears are made with the same care and provided with the same gauges and clamping bar as those shown on the opposite page. With our Power Squaring Shears we can furnish countershafts if so ordered.

No. 31XX.	Will cut $\frac{3}{16}$ Iron, 31 inches long, Geared 5 to 1, weighs 2700 lbs.,		$500 00
No. 42XX.	Will cut No. 10 Iron, 42 inches long, Geared 4 to 1, weighs 2200 lbs.,		400 00
No. 50XX.	Will cut No. 10 Iron, 50 inches long, Geared 4 to 1, weighs 2700 lbs.,		525 00
No. 72XX.	Will cut No. 10 Iron, 72 inches long, Geared 4 to 1, weighs 3000 lbs.,		600 00
Extra Blades for No. 31XX,		per pair,	15 00
Extra Blades for No, 42XX,		per pair,	25 00
Extra Blades for No. 50XX,		per pair,	35 00

SQUARING SHEARS.

Stow's Power Squaring Shears.

These Shears have been constructed with especial reference to workers in metals who do not desire as heavy or as expensive a Shear as illustrated on pages 78 and 79. They are fitted with front and back gauges, are made with great care, and are warranted to cut metal as thick as represented.

BACK-GEARED, WITH HOLD DOWN ATTACHMENT.

No. 020.	Will cut No. 16 Iron, 20 inches in length, ...	**$146 00**
No. 030.	Will cut No. 16 Iron, 30 inches in length, ...	**210 00**
No. 040.	Will cut No. 16 Iron, 42 inches in length, ...	**250 00**
No. 050.	Will cut No. 16 Iron, 50 inches in length, ...	**300 00**
No. 060.	Will cut No. 16 Iron, 60 inches in length, ...	**350 00**
No. 072.	Will cut No. 16 Iron, 72 inches in length, ...	**450 00**
No. 084.	Will cut No. 16 Iron, 84 inches in length, ...	**600 00**

These Shears may be ordered without Hold Down Attachment.

SQUARING SHEARS.

Stow's Power Squaring Shears.

These Shears are constructed in the same manner as those represented on page 80, with the exception that they are not back geared, therefore will not cut as heavy metal. We warrant them, however, to cut No. 18 Iron. They are arranged with our patent gauges, which are 30 inches in length and admit of rapid adjustment. The machines are well made; all parts are nicely fitted. No squaring shears manufactured give better satisfaction to the operator.

NOT BACK-GEARED, WITH HOLD DOWN ATTACHMENT.

No. AX20.	Will cut No. 18 Iron, 20 inches long, ...	**$90 00**
No. AX30.	Will cut No. 18 Iron, 30 inches long, ...	**150 00**
No. AX40.	Will cut No. 18 Iron, 42 inches long, ...	**200 00**
No. AX50.	Will cut No. 18 Iron, 50 inches long, ...	**250 00**
No. AX60.	Will cut No. 18 Iron, 60 inches long, ...	**300 00**
No. AX72.	Will cut No. 18 Iron, 72 inches long, ...	**400 00**

These Shears may be ordered without Hold Down Attachment.

SQUARING SHEARS.

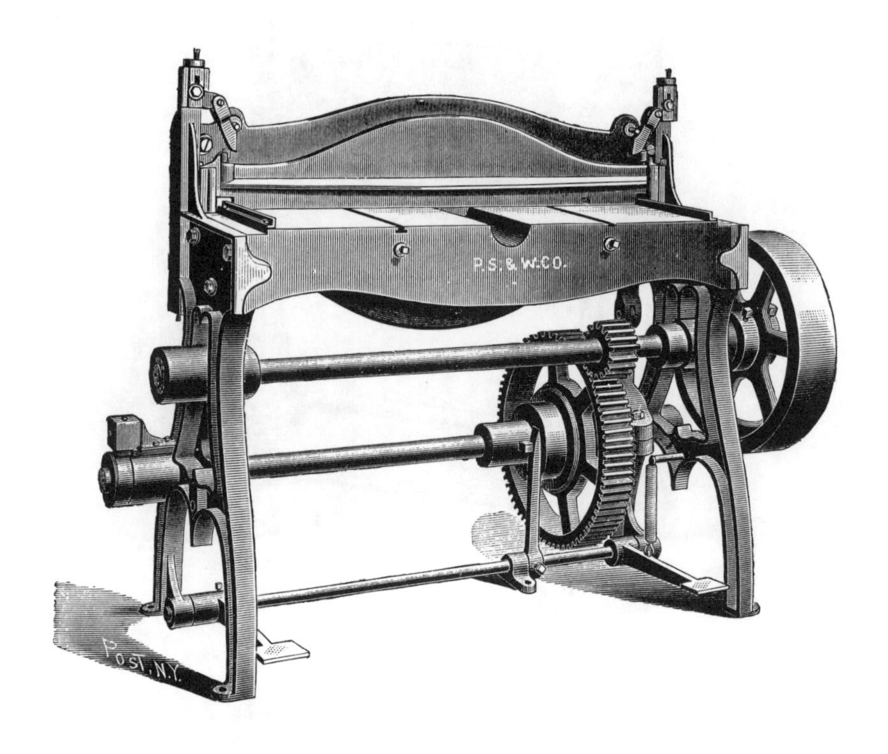

Stow's Power Squaring Shears.

The above cut represents a Shear of great power and durability. These Shears are fitted with our Patent Quick Acting Back Gauge, Front and Side Gauges and Side Extension Tables. The back gauge is easily and accurately set. They bear the same guarantee of excellence as other shears of our manufacture.

BACK-GEARED, WITH HOLD DOWN ATTACHMENT.

No. 1300.	Will cut No. 14 Iron, 31 inches long, weighs 1230 lbs., .	**$230 00**
No. 1360.	Will cut No. 14 Iron, 37 inches long, weighs 1325 lbs., . .	**265 00**
No. 1400.	Will cut No. 14 Iron, 42 inches long, weighs 1375 lbs., .	**300 00**

These Shears may be ordered without Hold Down Attachment.

SQUARING SHEARS.

Stow's Power Squaring Shears.

The above cut represents a Power Shear like the one on page 82, except that it is not back-geared and will not cut as heavy metal. Side extension tables are attached to these shears.

We make irregular shaped blades for cutting any desired shape, such as elbow sections, corset steels, saw blades, etc., to be used either with foot or power shears. Prices given on receipt of sample indicating shape or description of work.

Not Back-Geared, with Hold Down Attachment.

No. 2300.	Will cut No. 16 Iron, 31 inches long, weighs 672 lbs., .	**$190 00**
No. 2360.	Will cut No. 16 Iron, 37 inches long, weighs 690 lbs., . .	**210 00**
No. 2400.	Will cut No. 16 Iron, 42 inches long, weighs 825 lbs., .	**235 00**

These Shears may be ordered without Hold Down Attachment.

Stow's Shears for Cutting Corrugated Metal.

The above cut represents our Squaring Shears constructed for cutting corrugated metal. The blades are shaped to fit the corrugation to be cut. Blades for different shaped corrugations can be fitted to the same Shear. They are desirable for Cornice Makers, Roofers, etc. They are fitted with Back and Side Extension Tables and our Patent Gauges. The springs are now provided with a screw for regulating the tension and compensating for wear.

No. 1100.	Will cut No. 16 Iron, 31 inches long, weighs 460 lbs., . .	**$120 00**
No. 1200.	Will cut No. 16 Iron, 42 inches long, weighs 625 lbs., . .	**210 00**

SQUARING SHEARS.

Stow's Extra Heavy Squaring Shears.

These Shears are made from new patterns, with and without Hold Down Attachments, and with side extension tables. We unhesitatingly recommend these Shears as being made in the most thorough manner and warranted to cut metal as heavy as represented. All our Stow's Squaring Shears are fitted with our Patent Quick Acting Back Guage, which is easily and accurately set, and operated more quickly than any gauge manufactured. The side springs are fitted with a screw for regulating the tension and compensating for wear.

WITHOUT HOLD DOWN ATTACHMENT.

No. 0130.	Will cut No. 16 Iron, 31 inches long, weighs 550 lbs.,	$70 00
No. 0136.	Will cut No. 16 Iron, 37 inches long, weighs 675 lbs.,	90 00
No. 0140.	Will cut No. 16 Iron, 42 inches long, weighs 700 lbs.,	150 00

WITH HOLD DOWN ATTACHMENT.

The following are in all respects like the Shears illustrated above, except they are constructed with our Patent Automatic Hold Down Attachment.

No. 0300.	Will cut No. 16 Iron, 31 inches long, weighs 580 lbs.,	$75 00
No. 0360.	Will cut No. 16 Iron, 37 inches long, weighs 705 lbs.,	95 00
No. 0400.	Will cut No. 16 Iron, 42 inches long, weighs 740 lbs.,	160 00

STOW'S CURVED SHEARS.

Stow's Curved Shears are similar in general design to the one illustrated above, but are arranged for holding Curved Blades of special designs for cutting irregular shapes, such as Elbow Blanks, etc. They are accurately adjusted and perfect in every respect.

Prices furnished upon application with description or sample of work.

SQUARING SHEARS.

Stow's Squaring Shears.

The above cut represents our Improved Squaring Shears, with table, front extension arms and patent quick acting back gauge. The side springs are provided with a set screw arranged to regulate the tension and compensate for wear. The back gauge is simple, being constructed of tubing; the operator who wishes to lengthen the gauge can substitute other tubing of the desired length. The arms are 30 inches long. The adjustment is rapid and accurate. These shears are so constructed that side extension tables can be easily attached, as shown on page 87, but prices named below *do not include side extension tables*. The e shears are well made, well fitted, and thoroughly tested.

No. 14. Will cut No. 20 Iron, 14 inches long,		**$35 00**
No. 20. Will cut No. 20 Iron, 22 inches long,		**35 00**
No. 25. Will cut No. 20 Iron, 25 inches long,		**45 00**
No. 30. Will cut No. 18 Iron, 31 inches long,		**50 00**
No. 40. Will cut No. 18 Iron, 42 inches long,		**130 00**
No. 50. Will cut No. 18 Iron, 50 inches long,		**190 00**
No. 60. Will cut No. 18 Iron, 60 inches long,		**240 00**
No. 72. Will cut No. 18 Iron, 72 inches long,		**275 00**
Extra Blades for Nos. 14 or 20,	per pair,	**9 00**
Extra Blades for No. 25,	per pair,	**10 50**
Extra Blades for No. 30,	per pair,	**12 00**
Extra Blades for No. 40,	per pair,	**22 00**
Extra Blades for No. 50,	per pair,	**30 00**
Extra Blades for No. 60,	per pair,	**34 00**

SQUARING SHEARS.

Stow's Squaring Shears.

The illustration above represents our Squaring Shears as now made, with top and side extension tables and patent quick acting back gauge. They are also provided with front gauges and front extension arms as shown above. The springs are now arranged with a set screw for regulating the tension and compensating for wear. Every Squaring Shear, when it leaves our factory, will cut wet tissue paper the whole length of the blade. A wrench is furnished with each Shear. The wire cast in the leg furnishes a pocket in which the wrench should be kept, so as to be always with the Shear and ready for use.

No. 120. Will cut No. 20 Iron, 22 inches long,		$37 00
No. 130. Will cut No. 18 Iron, 31 inches long,		52 00
No. 136. Will cut No. 18 Iron, 37 inches long,		82 00
No. 140. Will cut No. 18 Iron, 42 inches long,		134 00
Extra Blades for No. 120,	per pair,	9 00
Extra Blades for No. 130,	per pair,	12 00
Extra Blades for No. 136,	per pair,	15 00
Extra Blades for No. 140,	per pair,	22 00

SQUARING SHEARS.

Stow's Squaring Shears, with Hold Down Attachment.

The above cut represents our newest and latest improved Squaring Shears, with top and side extension tables, patent extension arms, side springs and Hold Down Attachment.

This attachment is placed in front of the blades and gives a uniform pressure while the Shear is in operation, and holds the metal firm and keeps it from slipping. It can be adjusted to different thicknesses of metal by means of screws at the ends of the attachment. It operates automatically, and the pressure is secured by the perpendicular motion of the upper shear blade. It also shows our patent quick acting back gauge, which for ease of setting accurately is all that can be desired. The side springs are arranged with a screw for regulating the tension and compensating for wear.

No. 200.	Will cut No. 20 Iron, 22 inches long, weighs 370 lbs., . . .	**$40 00**
No. 300.	Will cut No. 18 Iron, 31 inches long, weighs 400 lbs., . .	**55 00**
No. 360.	Will cut No. 18 Iron, 37 inches long, weighs 700 lbs., ...	**85 00**
No. 400.	Will cut No. 18 Iron, 42 inches long, weighs 720 lbs., . .	**140 00**
No. 500.	Will cut No. 18 Iron, 50 inches long, weighs 800 lbs., . .	**200 00**
No. 600.	Will cut No. 18 Iron, 60 inches long, weighs 875 lbs., . .	**255 00**

SQUARING SHEARS.

Wilcox Pattern Squaring Shears.

No. 000, for Power only, will cut No. 12 Iron. Nos. 00 and 0 will cut No. 16 Iron. No. 1 will cut Sheet Iron and Tin. No. 2, Tin only.

Nos. 1 and 2 are arranged to work by hand lever or by foot power. They are much liked in many sections of the country. They are made with back and front gauges; and are for light work where much speed in execution is desired.

No. 000. With Iron Frame, will cut No. 12 Iron, 72 inches long, weighs 2000 lbs.,		**$800 00**
No. 00. With Iron Frame, will cut No. 16 Iron, 37 inches long, weighs 700 lbs.,		**140 00**
No. 0. With Iron Frame, will cut No. 16 Iron, 30 inches long, weighs 650 lbs.,		**125 00**
No. 1. With Iron Frame, will cut No. 20 Iron, 30 inches long, weighs 300 lbs.,		**55 00**
No. 1. Without Iron Frame, will cut No. 20 Iron, 30 in. long, weighs 250 lbs.,		**45 00**
No. 2. With Iron Frame, will cut Tin, 20 inches long, weighs 225 lbs.,		**38 00**
No. 2. Without Iron Frame, will cut Tin, 20 inches long, weighs 190 lbs.,		**32 00**
Extra Blades for No. 00,	per pair,	**20 00**
Extra Blades for No. 0,	per pair,	**16 00**
Extra Blades for No. 1,	per pair,	**12 00**
Extra Blades for No. 2,	per pair,	**9 00**

SQUARING SHEARS.

Hull's Acme Squaring Shears.

These shears are manufactured at our Works in CLEVELAND, OHIO, and are shipped from that point. They are arranged with Table, Back and Front Gauges, for squaring, stripping and cutting at any desired angle without necessity of marking the sheets.

The six and eight foot Squaring Shears have the back gauges arranged so that they are moved by a crank at one end of the machine.

NOTE.—In ordering these Shears, be careful to give NAME as well as NUMBER, so as not to confound with the WILCOX SQUARING SHEARS represented on page 89, which bear the same numbers.

No. 0.	Acme, will cut Tin, 22 inches long,	**$35 00**
No. 1.	Acme, will cut Tin, 25 inches long,	**45 00**
No. 2.	Acme, will cut No. 18 Iron, 31 inches long,	**50 00**
No. 2½.	Acme, will cut No. 18 Iron, 37 inches long,	**80 00**
No. 3.	Acme, will cut No. 16 Iron, 42 inches long,	**130 00**
No. 4.	Acme, will cut No. 16 Iron, 52 inches long,	**190 00**
No. 5.	Acme, will cut No. 18 Iron, 61 inches long,	**240 00**
No. 6.	Acme, will cut No. 18 Iron, 72 inches long,	**275 00**
No. 7.	Acme, will cut No. 20 Iron, 86 inches long,	**325 00**
No. 8.	Acme, will cut No. 20 Iron, 98 inches long,	**375 00**

Extra Blades as on page 86.

SQUARING SHEARS.

Hull's Acme Squaring Shears.

Our Acme Shears are deservedly popular throughout the West on account of their many excellences. The parts are accurately fitted, and they are made in the most thorough and substantial manner by skilled mechanics. Those described on this page are arranged with top and side extension tables, improved front and back gauges, and include our latest improvements.

These Shears are designed for *cutting metal*, but are adapted to a variety of *uses* and *special work*, as *cutting leather, cloth, paper, hard or soft rubber, etc.*

All these Shears can be arranged for power, as shown on previous pages; back-geared or not back-geared, with or without hold down attachments. They also can be fitted with the lever arc automatic gauge as used on our Cornice Makers' Shear, described on page 92.

No. 730. Acme, will cut No. 18 Iron, 31 inches long,		**$52 00**
No. 731. Acme, will cut No. 14 Iron, 31 inches long,		**80 00**
No. 736. Acme, will cut No. 18 Iron, 37 inches long,		**82 00**
No. 740. Acme, will cut No. 16 Iron, 42 inches long,		**134 00**
Extra Blades for No. 730,	per pair,	**12 00**
Extra Blades for No. 736,	per pair,	**15 00**
Extra Blades for No. 740,	per pair,	**22 00**

SQUARING SHEAR.

Cornice Makers' Squaring Shear.

WITH LEVER ARC AUTOMATIC GAUGE.

In addition to the large line of Squaring Shears already described we have recently manufactured a special Shear to meet the wants of cornice makers and others requiring a first-class Shear of great capacity, and embodying the latest improvements and at a low price. As shown in the above illustration it has an automatic HOLD DOWN attachment, gauges for squaring and cutting at angles, and our NEW DEVICE for working the back gauge with gear and rack by means of lever in place of bevel gears and screws as heretofore used. The lever is operated from the centre of the Shear on a graduated arc which indicates the position of the gauge. The lever is operated from the front instead of the back of the Shear. To cut sheets 1 to 11 inches wide simply move the lever to the corresponding measurement indicated on the arc. To cut from 11 to 20 inches in width, move the gauge to the second pocket in the back arms and operate the lever in the same manner as in cutting narrow widths. We manufacture this Shear only in one size, and guarantee it to cut No. 22 Iron the entire length of the blades.

No. 960. Lever Arc Shear, will cut No. 22 Iron, 97 inches long, weighs 2400 lbs., **$250 00**

SLITTING OR SQUARING SHEARS.

Continuous Cutting or Gap Shears.

WITH HOLD DOWN ATTACHMENT.

These Shears are made from entirely new patterns, and are arranged with gauges for cutting sheet metal of any length into strips from ½ inch to 15 inches in width. Sheet metal 37 inches long may be cross cut at any point desired.

To cut metal into strips longer than 37 inches, the gauges at the end of the shears should be removed. No other adjustment is necessary.

The hold down attachment is operated by hand and slides on independent ways, so that when the blades are in motion the hold down remains in a fixed position. It can be operated from either end of the shear by means of the small hand wheels. It is arranged with eccentrics, so that when the hold down is brought against the plate it holds itself firmly in position.

They are fitted with side extension tables for supporting long sheets of metal.

Distinct lines are planed in the bed plate parallel with the blades, to aid in adjusting the gauges.

Adjustable gauges are provided for the side extension tables, to be used when cutting long sheets.

The treadle is arranged so that the connecting rod can be instantly changed to give increased leverage for cutting heavy metal. The No. 85 shear is now furnished with our extension treadle, not shown in the illustration. This extension will be found useful when cutting metal heavier than No. 18 gauge.

No. 85.	For foot power, 37 inches, will cut No. 16 Iron, weighs 1100 lbs., .	**$125 00**
No. 185.	For power, 37 inches, will cut No. 16 Iron, weighs 1600 lbs., . .	**225 00**

BENCH AND SLITTING SHEAR.

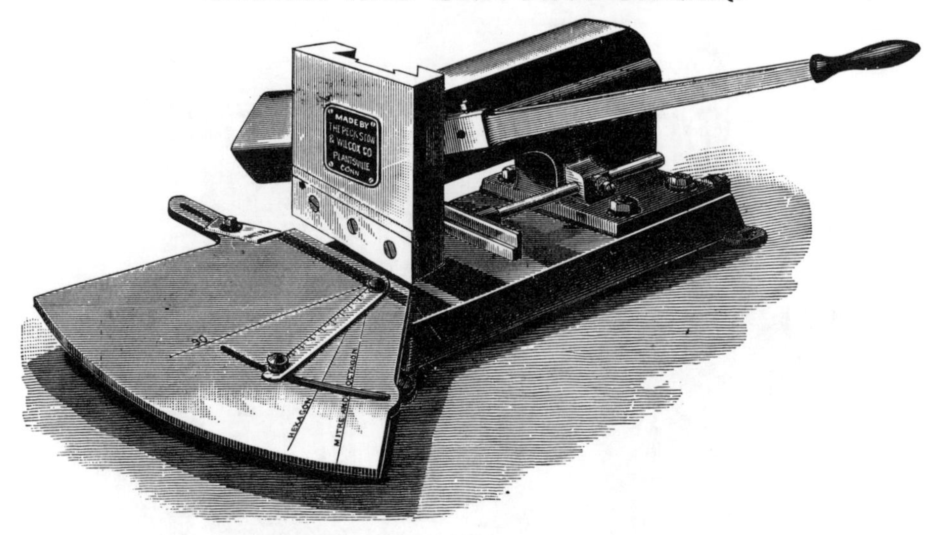

Combined Bench and Slitting Shear.

LENGTH OF BLADE 6 INCHES.

This Shear is constructed on entirely new principles. It surpasses all other Shears in the variety of work performed. Its superiority over other shears in sheet metal cutting is as follows:

First.—It will do the work of an ordinary bench shear, over which it has the following advantages: The length of cut is longer to the same movement of hand: the same pressure of hand will cut thicker stock; it cuts with the same ease at all points of the cut, while an ordinary bench shear cuts harder near the point than near the bolt.

Second.—The lower blade of this shear is stationary, so that when cutting to line the mark may easily be followed with accuracy. The blades are so constructed that the line drawn is always exposed to the view of the operator.

Third.—The lift of the upper blade is adjusted to any desired height by a screw in the lever. When so adjusted as to permit the sheet to be passed through under the blades the rear gauge may be so set as to cut slits, or square, octagonal, hexagonal, or any other shaped hole, with straight sides, in the centre of a sheet of metal 20 inches square or less.

Fourth.—It will cut round or elliptical bottoms for vessels of any size having a radius of 2 inches or more.

Fifth.—The front table can be removed at pleasure, and should be removed when cutting to line. By placing this table in position as seen in the cut, and fastening gauge in proper place, roofing plates can be cornered ready for turning the lock. This table is so graduated that at a glance the gauges can be set at any desired angle between 45 degrees and 90 degrees.

Sixth.—The blades can be easily removed and ground. The shear is so constructed that it can be readily adjusted, and with proper care should last a lifetime. When used as an ordinary bench shear the table should be removed.

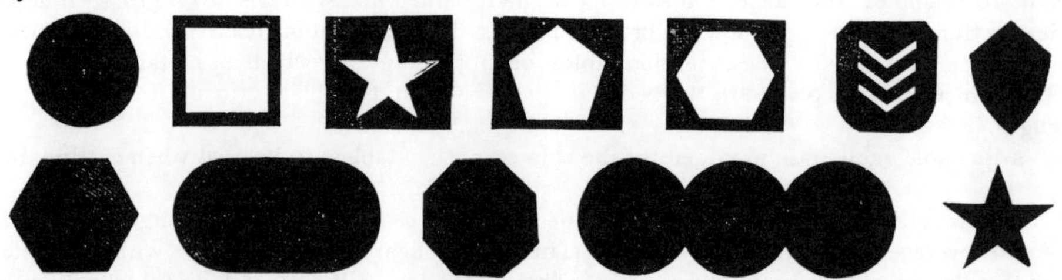

The above cut shows some of the shapes which can be cut with this shear from the centre of a sheet of plate metal.

No. 7. Bench and Slitting Shear, will cut No. 16 Iron, weighs 132 lbs., . . . **$25 00**
Extra Blades, ... per pair, **3 50**

BENCH AND SLITTING SHEAR.

Combined Bench and Slitting Shear.

LENGTH OF BLADE 9 INCHES.

This Shear is constructed on the same plan as our No. 7 Slitting Shear, represented on the opposite page, except it stands on legs as shown in the cut. This Shear will cut sheet metal as thick as No. 12 gauge. The length of reach on to a sheet of thick metal is 15½ inches; on to a sheet of thin metal, 19 inches.

To cut heavy iron, or thicker than No. 18, the stay bolt should always be in its place; when cutting No. 18 iron, or thinner, it may be removed to give it more reach. It has a compound lever attachment to be used to increase its power when required; this should not be used when the stay bolt is removed.

This Shear will be found very serviceable for *Range Manufacturers* and others who require a shear for cutting oven doors and odd shapes for special designs.

No. 9. Bench and Slitting Shear, will cut No. 12 Iron, weighs 650 lbs., . .		**$150 00**
Extra Blades, ...	per pair,	**4 00**

ROTARY SLITTING SHEARS.

Slitting Shear, with Gauge Table.

This Slitting Shear is designed for cutting thin metal into strips from ½ to 9 inches in width. It is arranged with gauges and a supporting table, so that accuracy may be secured.

No. 20. Rotary Slitting Shear, will cut No. 20 Iron, with Stand, weighs 60 lbs.,		**$25 00**
Extra Cutters, ...	per pair,	**6 00**

Power Slitting Shears, with Supporting Arm.

The above illustration represents a strong, heavy Shear for cutting brass, copper, sheet iron and other metals into strips of any length. They are arranged with gauges easily adjusted to cut different widths up to 17 inches.

No. 12. Back geared, will cut No. 12 Iron, weighs 600 lbs., ...		**$150 00**
No. 18. Not back-geared, will cut No. 18 Iron, weighs 425 lbs., ...		**120 00**
Extra Cutters, ...	per pair,	**18 00**

BEVELING AND PARALLEL SHEARS.

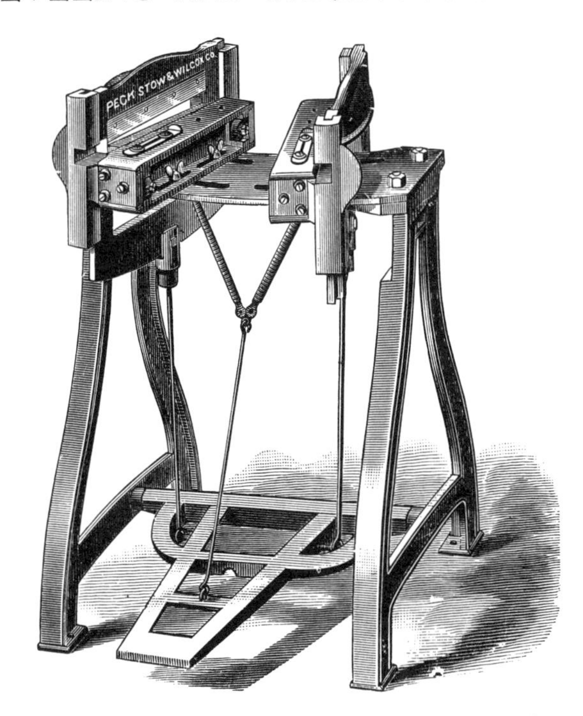

Beveling Shears, for Cutting the Ends of Pan Sections.

The above cut represents Lowe's Patent Beveling Shear, designed for cutting at one operation and at any desired angle the ends of the section, after having been cut by the parallel shears. No 1 will cut segments of circles 10 inches wide or less, and in length from 8 to 18 inches. No. 2 will cut segments of circles 12 inches wide or less, and in length from 5 to 24 inches. No. 3 will cut segments of circles 20 inches wide or less, and in length from 5 to 26 inches.

No. 1.	Lowe's Beveling Shears, 10 inch Blades, weighs 215 lbs., ...	**$45 00**
No. 2.	Lowe's Beveling Shears, 12 inch Blades, weighs 245 lbs., . . .	**55 00**
No 3.	Lowe's Beveling Shears, 20 inch Blades, weighs 325 lbs., ...	**75 00**

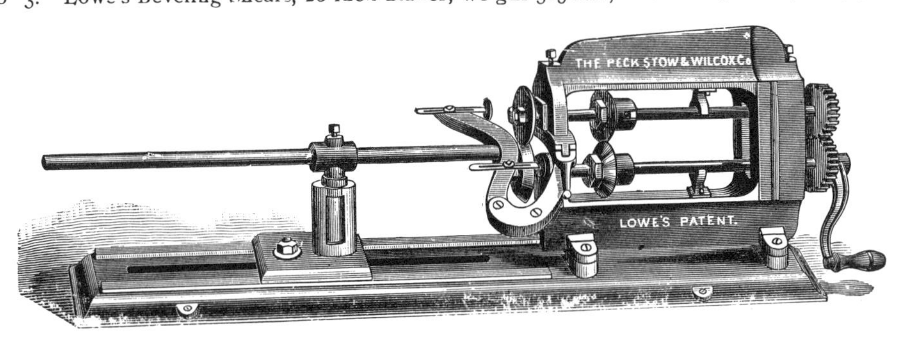

Lowe's Patent Shears for Cutting Parallel Curves.

The above cut represents a Machine designed for cutting from sheet metal parallel curves at one operation such as pan sections etc.

They can be readily adjusted for cutting parallel curves of any ordinary flaring vessel. They do the same work as is ordinarily done by the aid of dies in large establishments

No. 1 will cut the segments of circles from 4½ feet in diameter down to 9 inches, with a range in width from 12 inches to 3¾ inches.

No. 2 will cut the segments of circles from 9 feet in diameter down to 9 inches, with the same range in width as No. 1.

Attachments for cutting circles can be furnished for these machines at an extra cost varying according to size.

These Machines can be constructed for power at an additional expense of $15.00, without countershafts.

These Machines will not cut metal heavier than XXXX tin plate.

No. 1.	Lowe's Parallel Shear, weighs 100 lbs., ...	**$80 00**
No. 2.	Lowe's Parallel Shear, weighs 160 lbs., ...	**100 00**

ROTARY SHEARS.

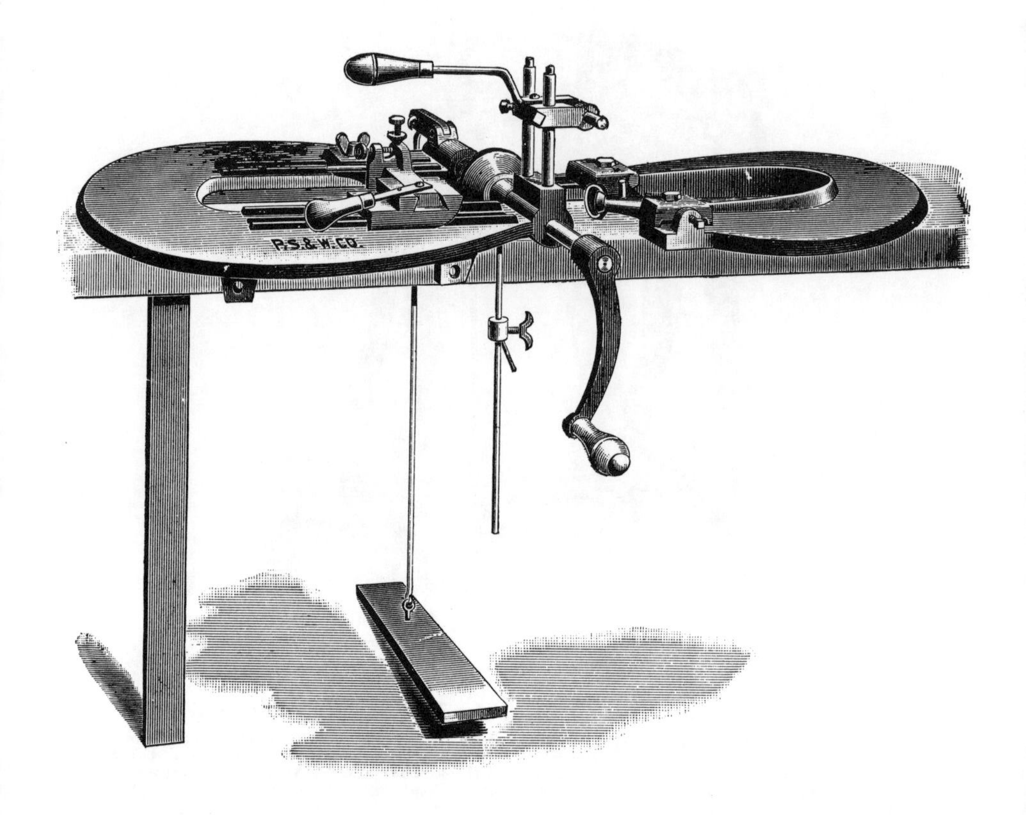

Savage's Patent Combination Circular Shear.

NO. 1, WITH FOUR PAIRS DISCS—ONE PAIR EACH OF 2⅛, 3, 4⅞ AND 7⅝ INCHES DIAMETER.

These Shears will cut Circles and bend or burr the same at any desired angle without extra discs. No. 1 will cut, turn the edge and burr circles from 2⅛ to 23 inches in diameter from XX tin. The burring attachment turns an edge on any circle cut, without extra discs, at a little greater angle than a right angle.

No. 1. For Light Metal, with Burring Attachment, weighs 150 lbs.,		**$35 00**
No. 1. For Light Metal, without Burring Attachment, weighs 140 lbs.,		**30 00**
No. 1. With both Burring Attachment and Edge Turner, weighs 155 lbs.,		**40 00**
Extra Discs for No. 1, not over 8 inches in diameter,	per pair,	**1 50**
Extra Cutters for No. 1,	per pair,	**1 50**
Extra Cutter Stocks for No. 1,	per pair,	**2 50**
Extra Edge Turner for No. 1,		**5 00**
Extra Burring Attachment for No. 1,		**5 00**

ROTARY SHEARS.

Savage's Patent Combination Circular Shears.

Nos. 00 and 0, with One Pair Discs.

The Shears above represented are made so they can be used by hand or power, but are without Burring attachment. They have an adjusting screw, and a scale on the bed, so that a very accurate adjustment can be made.

No. 00 will cut circles from 6 to 43 inches in diameter from No. 14 iron.
No. 0 will cut circles from 5 to 30 inches in diameter from No. 20 iron.
No. 00 has pulley 14 x 5¼ inches.
No. 0 has pulley 13 x 3⅜ inches.

No. 00. With one pair discs, not over 12 inches, weighs 1200 lbs., .		**$500 00**
No. 0. With one pair discs, not over 12 inches, weighs 555 lbs., . .		**300 00**
Extra Cutters for Nos. 00 and 0, ...	per pair,	**3 50**
Extra Discs for Nos. 00 and 0, not over 15 inches, . . .	per pair,	**9 00**

ROTARY SHEARS.

Newton's Patent Circular Shear.

WITH TWO PAIRS DISCS—ONE PAIR EACH OF 2⅜ AND 5⅞ INCHES IN DIAMETER.

Will cut Circles from 3 to 14 inches in diameter from Tin.

Newton's Circular Shear, weighs 125 lbs., ... , . . . **$22 00**

Extra Discs or Cutters, ... per pair, **2 00**

Extra Sliding Gauge, ... **1 50**

ROTARY SHEARS.

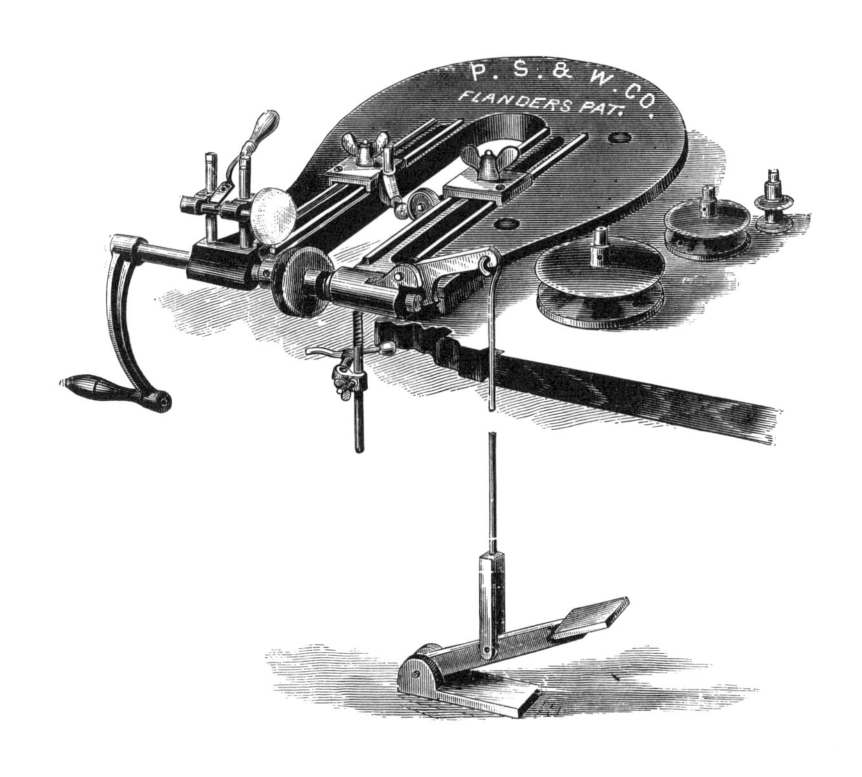

Flander's Patent Circular Shears.

WITH FOUR PAIRS OF DISCS—ONE PAIR EACH OF 1⅞, 3, 4⅞ AND 7⅝ INCHES IN DIAMETER.

No. 1 will cut Circles from 2½ to 23 inches in diameter.

No. 1. For Hand, for Tin, etc., with four pairs Discs, one pair Cutters and Edge Turner, weighs 110 lbs.,		**$30 00**
No. 1. Without Edge Turner, weighs 100 lbs,		**25 00**
Extra Discs for No. 1, not over 8 inches in diameter,	per pair,	**1 50**
Extra Cutters for No. 1,	per pair,	**1 50**
Extra Cutter Stocks for No. 1,	per pair,	**2 50**
Extra Edge Turner for No. 1,		**5 00**

ROTARY SHEARS.

Waugh's Patent Bevel, Square and Circular Shears.

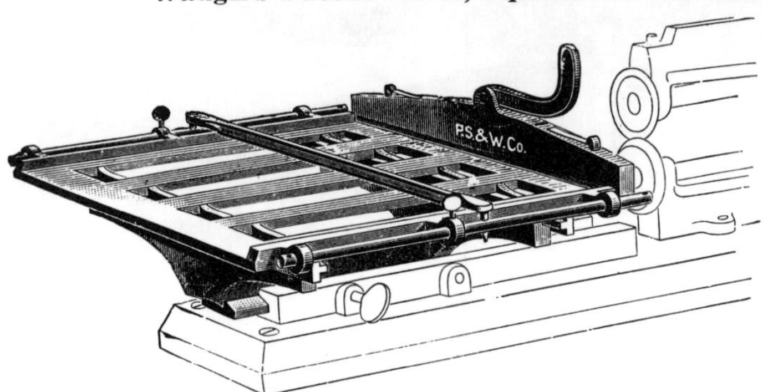

Squaring Attachment.

These Shears are widely known, and have won an enviable reputation. The cutters are always set and ready for use. Each machine includes a rim gauge for cutting rims or cover hoops, bevel or straight.

Nos. 1 and 2 will cut any square or bevel under 20 inches.

No. 3 is made for both hand and power, and may be ordered with a 20 or 30 inch bevel and squaring attachment.

No. 4, for power only, is back geared; has a 12 inch throat and 4 inch cutters.

No. 5, for power, is back geared 4 to 1; has a 9 inch throat and 5 inch cutters.

No. 1.	Will cut Circles from tin, 2¼ to 15 inches, weighs 60 lbs.,	$33 00
No. 1.	With Bevel and Squaring Attachment, weighs 70 lbs.,	43 00
No. 2.	Will cut Circles from tin, 2¼ to 20 inches, weighs 70 lbs.,	38 00
No. 2.	With Bevel and Squaring Attachment, weighs 80 lbs.,	48 00
No. 3.	Will cut Circles from No. 20 Iron, 4½ to 40 inches, weighs 225 lbs.,	75 00
No. 3.	With 20 inch Bevel and Squaring Attachment, weighs 240 lbs.,	90 00
No. 3.	With 30 inch Bevel and Squaring Attachment, weighs 250 lbs.,	100 00
No. 4.	Will cut Circles from No. 18 Iron, 4 to 48 inches, weighs 500 lbs ,	150 00
No. 5.	Will cut Circles from No. 14 Iron, 5 to 30 inches, weighs 650 lbs.,	175 00

COMBINED RING AND CIRCULAR SHEAR.

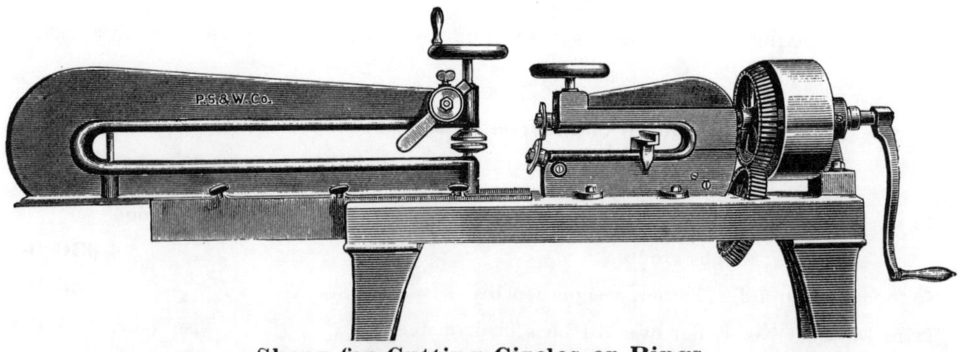

Shear for Cutting Circles or Rings.

This Machine is constructed so that it can be used either by hand or power. It is designed for cutting rings from a sheet of metal without cutting through the outer edge, that is, internal circles; it also can be used as a regular Circular Shear. It is made in one size only, and will cut circles from 3 to 40 inches in diameter, and internal rings from 3 to 39 inches in diameter. It can be used on Sheet Iron as heavy as No. 16, It is provided with both pulley and crank.

No. 20. Ring and Circular Shear, on Legs, weighs 400 lbs., ... **$120 00**

SHEAR AND PUNCH.

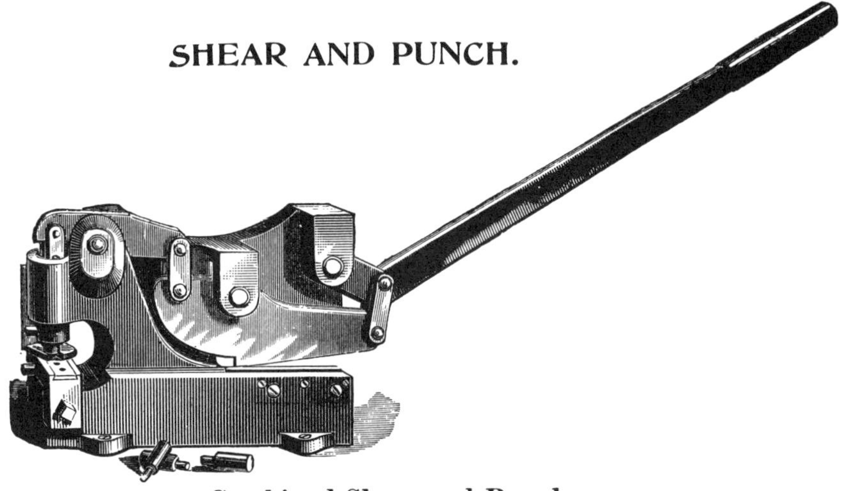

Combined Shear and Punch.

These Machines are suitable for both cutting and punching metal. They may be used for cutting $\frac{1}{8}$ inch iron, and $\frac{3}{16}$ and $\frac{1}{4}$ inch narrow bars. Four Punches are put up with No. 1, viz., $\frac{1}{8}$, $\frac{3}{16}$, $\frac{1}{4}$ and $\frac{5}{16}$ inch. Five Punches are put up with No. 2, viz., $\frac{1}{8}$, $\frac{3}{16}$, $\frac{1}{4}$, $\frac{5}{16}$ and $\frac{3}{8}$ inch.

No. 1.	Shear and Punch, $4\frac{1}{2}$ inch cut, will punch $\frac{1}{8}$ inch iron, weighs 100 lbs.,	**$16 00**
No. 2.	Shear and Punch, $5\frac{1}{2}$ inch cut, will punch $\frac{3}{16}$ inch iron, weighs 120 lbs.,	**20 00**

SAMSON PUNCH.

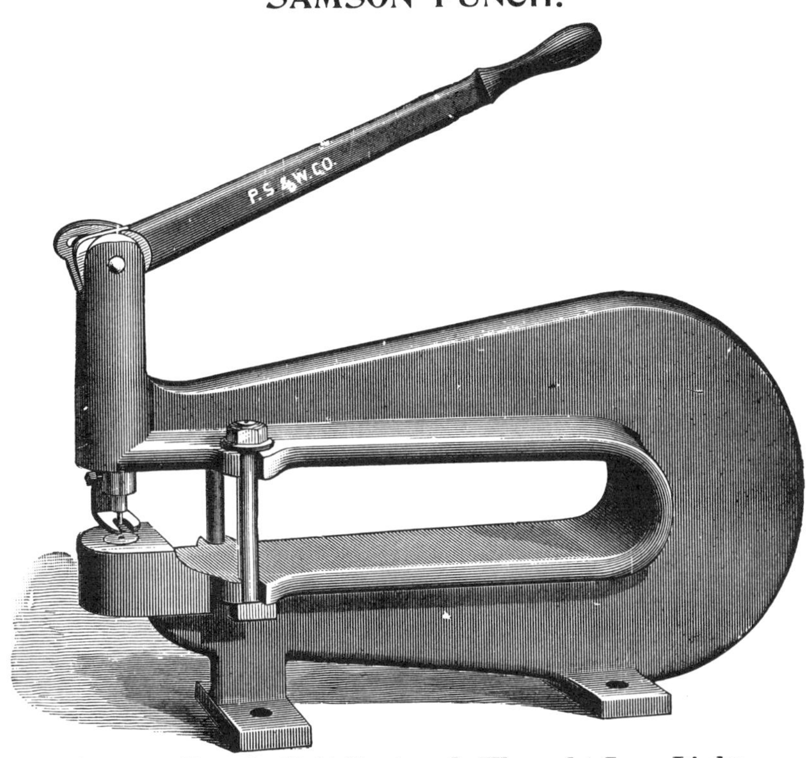

Samson Punch, Bolt Fastened, Wrought Iron Links.

We warrant it to punch No. 9 iron with "stay bolts," and No. 12 without "stay bolts." Depth of gap, 15 inches; will centre on 30-inch sheets without "stay bolts," and 7-inch sheets with "stay bolts." Accompanying each Machine are three sets of Dies and Punches, viz., $\frac{1}{8}$, $\frac{3}{16}$ and $\frac{1}{4}$ inch.

No. 5. Improved Samson Punch, weighs 95 lbs., ...		**$20 00**
Extra Punches, for Shear and Punch, or Samson Punch, ...	each,	**75**
Extra Dies, for Shear and Punch, or Samson Punch, ...	each,	**1 25**

LEVER SHEARS.

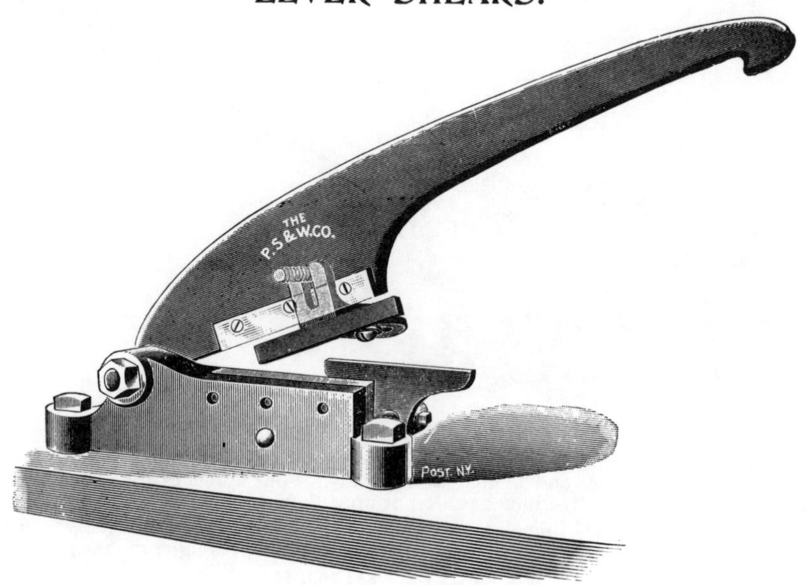

Improved Lever Shears, with Gauge.

These Shears are of extra weight, well made, and arranged with guages for cutting No. 18 iron.

No. 1.	Lever Shear, cuts 4½ inches, weighs 19 lbs., ...	$12 00
No. 1½.	Lever Shear, cuts 6 inches, weighs 24 lbs., ...	13 00
No. 2.	Lever Shear, cuts 8 inches, weighs 32 lbs., ...	15 00

NOTCHING MACHINE.

This Machine is adapted for notching the sections of pieced sheet metal ware, for cutting corners and hinge notches for square boxes, and other similar work. It will cut right angle corners 1 inch wide, 2 inches long, or smaller.

It is operated by a treadle, and will cut through several thicknesses of tin at one stroke. The gauges are adjustable; the die is made in sections, so that when dull they can be easily ground and reset; it is adjustable in all ways to compensate for its wear. It is better adapted to above uses than any machine we know of. It is compact, strong and easily operated.

In use always keep the Machine well oiled to keep the punch from grinding.

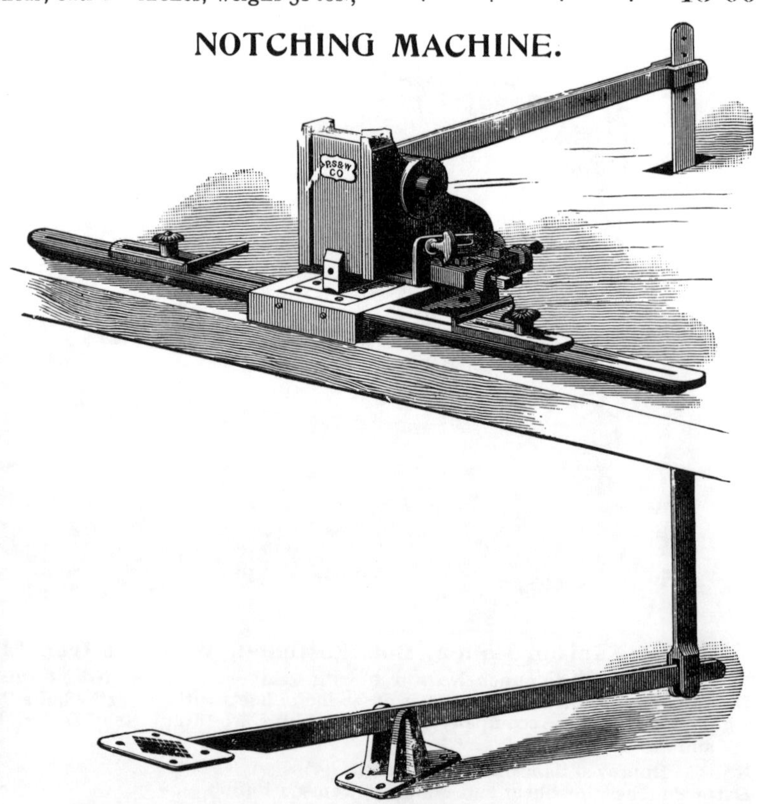

The Eureka Notching Machine.

Eureka Notching Machine, weighs 115 lbs., ... **$30 00**

ROLLING CUTTER SHEARS.

Continuous Rolling Cutter Shears.

BUCHER'S PATENT, OCT. 19, 1897.

This shear is operated by a pinion lever engaging in a rack and cutting with the forward roll of the cutting disc.

The advantage of this construction is that the power travels with the cutter. With ordinary shears as the cutting edge recedes from the fulcrum the resistance increases; not so with this shear, the resistance and power being the same all along the cut.

The motion of the circular cutting disc is about one-third of its circumference. The cutting disc is provided with three slots, which correspond with three pins on the cutting disc bolt; this enables the operator to change the cutting disc to six different positions; three on each side—the disc being reversible.

The lower or stationary straight cutter blade has two cutting edges, the ends of the cutter are beveled to match the heads of the bolts which draw the cutter blade in recess of shear stock. Two set screws which press this cutter blade in opposite directions are the means to adjust the lower cutter perfectly to the upper cutter disc.

The angle of contact of the two cutters remains the same during action, therefore this shear is well adapted to cut pieces out of a sheet of metal without injuring the same by cross-cut marks, which is very difficult with other shears.

The circular cutter on the front of the vertical plate or shear stock, and the pinion lever on the back of said plate, whose surfaces are perfectly parallel, when drawn up reasonably well by the nut of the bolt or shaft, bear so well against the plate that the friction of the two cutters can be totally obliterated by proper adjustment with the set screws in combination with the bevel head bolts, thus securing long service before any sharpening is necessary. This is surely the most simple and perfect shearing device for cutting sheet metal ever placed before the trade. It is especially adapted to tin and sheet metal workers in general.

No. 11.	Will cut No. 20 iron,	each,	**$4 00**
No. 12.	Will cut No. 16 iron,	each,	**6 50**

Extra Parts.

	No. 11.	No. 12.
Bed,	**$1 75**	**$3 00**
Disc,	**1 50**	**2 50**
Lower Blade,	**50**	**75**
Lever Complete,	**75**	**1 25**
Disc Bolt Complete,	**1 25**	**1 50**
Blade Clamp Bolts, each,	**20**	**20**

HAND CRIMPERS.

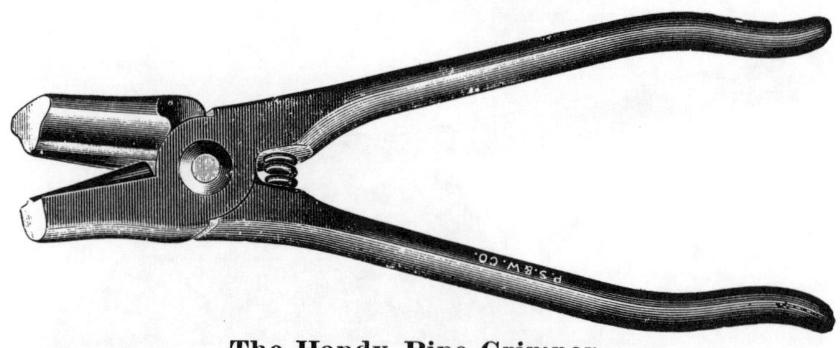

The Handy Pipe Crimper.

The Handy Pipe Crimper is intended principally to make a V-shaped crimp in the corners of square and oblong heater pipe, to admit of the crimped end entering the uncrimped. It will also work well on round and oval pipe.

It is a convenient and very practical crimper, and fills a long felt want for a low-priced hand crimper for shop use and to be sent out on jobs.

The Handy Pipe Crimper, ... per dozen, **$6 00**

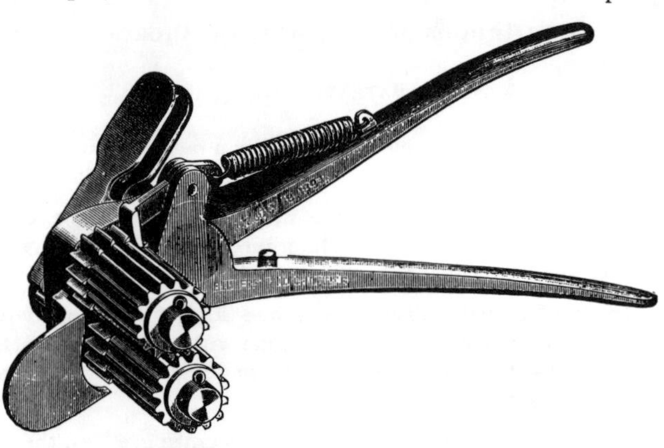

Rotary Hand Crimper.

This Rotary Hand Crimper will make a 1 inch crimp on metal as heavy as No. 24 gauge. The working parts are made of steel and any part can be duplicated.

Rotary Hand Crimper, ... each, **$2 00**

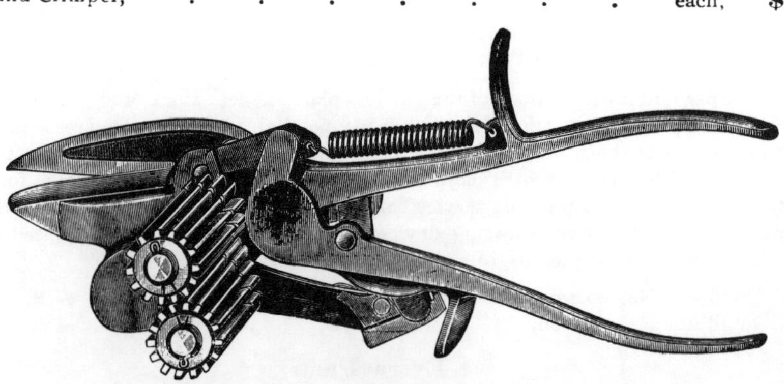

Combination Pipe Fitter.

The combination of a Shear and Rotary Hand Crimper will be found serviceable in sheet metal work of all kinds. It will make a 1 inch crimp on metal as heavy as No. 24 gauge, and will cut metal as heavy as No. 18 gauge.

All working parts are made of steel and duplicate parts can be furnished.

Combination Pipe Fitter, ... each, **$3 50**

TINNERS' SHEARS.

Bench Shears.

No. 00. Bench, cut 12 inches,	each,	$13 50
No. 0. Bench, cut 10½ inches,	each,	12 00
No. 1. Bench, cut 9 inches,	each,	8 00
No. 2. Bench, cut 8⅝ inches,	each,	7 00
No. 3. Bench, cut 8⅜ inches,	each,	6 00
No. 4. Bench, cut 8 inches,	each,	5 00
No. 5. Bench, cut 7 inches,	each,	4 00
No. 6. Bench, cut 6 inches,	each,	3 50
Elbow Bench, cut 4 inches,	each,	5 25
Elbow Bench, Extra Heavy, cut 6 inches,	each,	12 00
Elbow Bench, Double Extra Heavy, cut 7½ inches, will cut No. 12 Iron,	each,	25 00

TINNERS' HAND SHEARS OR SNIPS.

Hand Shears or Snips, Left Hand.

We can furnish any of the following Snips, with bows made for left-handed men, at an extra cost of *fifty cents each, net*. This means a right-hand cut with bows made for the left hand.

We sometimes have call for Snips for left-handed men, but a left-hand Snip, as we apply the term, is not intended for a left-handed man.

Confusion often arises as to the meaning of the terms RIGHT AND LEFT HAND as applied to Snips. Right hand are made like Bench Shears and left hand opposite. That is, when held in the right hand the lower blade is on the right side of the Shears in *right-hand Shears*, and on the left hand in *left-hand Shears*. Left hand only are shown in the cuts. Left-hand Snips are more generally used, and are invariably sent unless right hand are ordered.

All Snips stamped "P. S. & W. Co." are of the highest grade of excellence and fully warranted.

No. 06½. Hand, cut 4½ inches, extra heavy, entire length 18 inches,	each,	$4 50
No. 6½. Hand, cut 4½ inches,	each,	3 00
No. 7. Hand, cut 4 inches,	each,	2 50
No. 8. Hand, cut 3½ inches,	each,	2 00
No. 9. Hand, cut 3 inches,	each,	1 50
No. 10. Hand, cut 2½ inches,	each,	1 40

TINNERS' HAND SHEARS OR SNIPS.

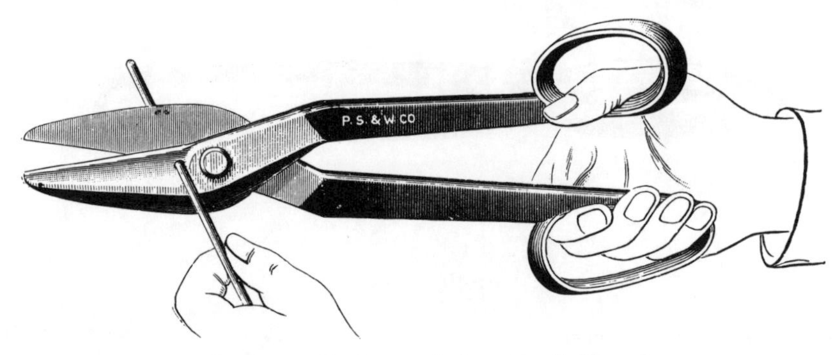

German Pattern Snips, Left Hand.

These Snips are made with a wire cutter, as shown in the illustration above.

No. 1.	Hand, cut 4½ inches,	each,	**$3 00**
No. 2.	Hand, cut 4 inches,	each,	**2 50**
No. 3.	Hand, cut 3½ inches,	each,	**2 00**
No. 4.	Hand, cut 3¼ inches,	each,	**1 75**
No. 5.	Hand, cut 3 inches,	each,	**1 50**
No. 6.	Hand, cut 2½ inches,	each,	**1 40**

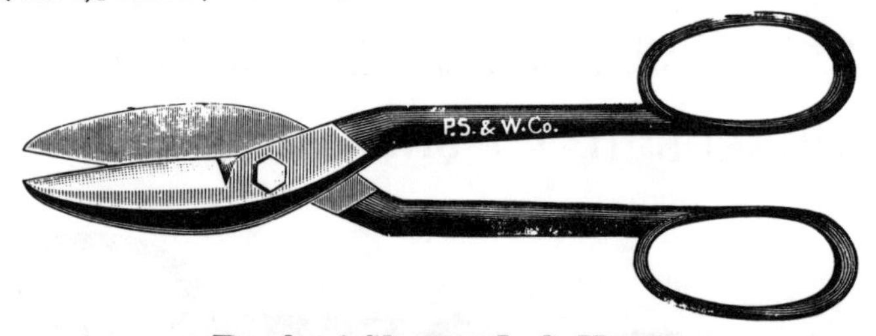

Roofers' Shears, Left Hand.

These Shears are especially adapted for roofing work, where lightness as well as strength is desirable. They are made with best cast steel blades and drop forged handles, and are fully guaranteed.

No. 70.	Hand, cut 4 inches,	each,	**$2 50**
No. 80.	Hand, cut 3½ inches,	each,	**2 00**
No. 90.	Hand, cut 3 inches,	each,	**1 50**
No. 100.	Hand, cut 2½ inches,	each,	**1 40**

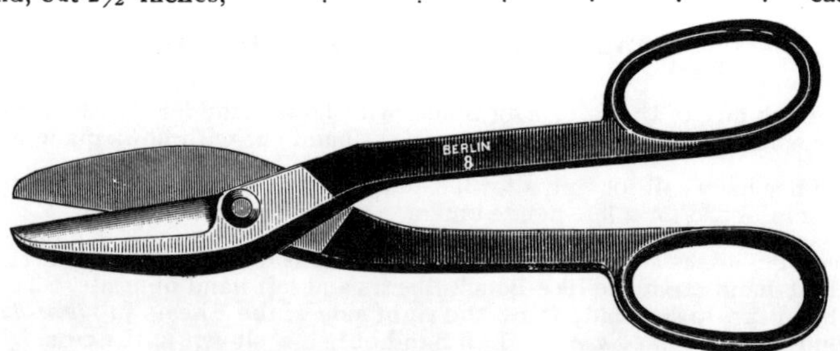

Hand Shears or Snips.—"Berlin."

At the solicitation of many valued customers we put upon the market a Hand Shear or Snip which we can warrant to be well made and serviceable for workers in sheet iron and metal as well as for general use. These Shears will bear our distinctive trade mark "BERLIN."

No. 46.	Hand, cut 4½ inches,	each,	**$3 00**
No. 47.	Hand, cut 4 inches,	each,	**2 50**
No. 48.	Hand, cut 3½ inches,	each,	**2 00**
No. 49.	Hand, cut 3 inches,	each,	**1 50**
No. 50.	Hand, cut 2½ inches,	each,	**1 40**

TINNERS' HAND SHEARS OR SNIPS.

Circular Hand Shears.—"P. S. & W Co."

These Circular Snips are our best quality, and of the highest grade of excellence. They are stamped with our name, "P. S. & W. Co.," and each pair is fully warranted. The experience of a century has enabled us to make the most perfect cutting tools, and no imitators have ever equalled us in the superior quality and finish of Tinners' Snips.

No. 7.	Circular, Hand,	each,	$3 50
No. 8.	Circular, Hand,	each,	3 00
No. 9.	Circular, Hand,	each,	2 50
No. 10.	Circular, Hand,	each,	2 25

Circular Hand Shears.—"Berlin."

Our Circular Snips stamped "Berlin" correspond in quality and finish to our regular "Berlin" Snips, represented and described on the opposite page.

No. 47.	Circular, Hand,	each,	$3 50
No. 48.	Circular, Hand,	each,	3 00
No. 49.	Circular, Hand,	each,	2 50
No. 50.	Circular, Hand,	each,	2 25

HAWK'S BILL SHEARS.

Improved Pointed Curved Hand Shears.

The above cut represents a Curved Shear of *real worth* and great merit It is capable of cutting in sheet metal, openings of any kind and shape. Letters are easily cut out from sheet metal. They are especially adapted for cutting off the bottoms of metal vessels, and for cutting openings in pipes or cylinders of every description, for furnace jackets, thimbles, tee joints, etc. A bottom can be cut from a pint cup or a copper boiler with equal ease.

No. 15.	Pointed Snips,	each,	$3 00

TINNERS' HAND SHEARS OR SNIPS.

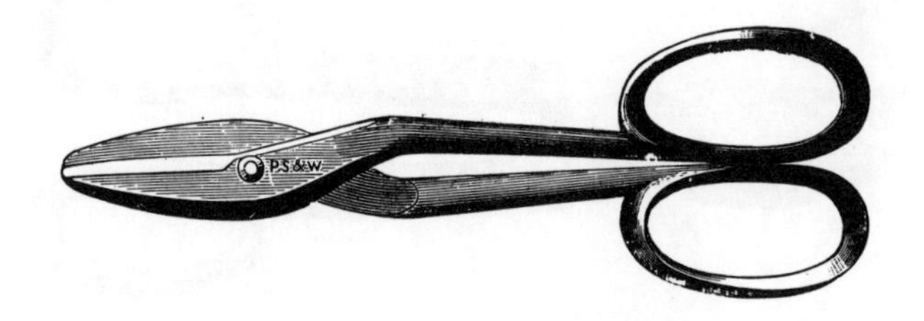

Jewelers' Snip, Solid Steel.

A Shear of fine finish and high grade, especially adapted to the use of Mechanics.

No. 11.	Hand, cut 2 inches, ...	each,	$1 25

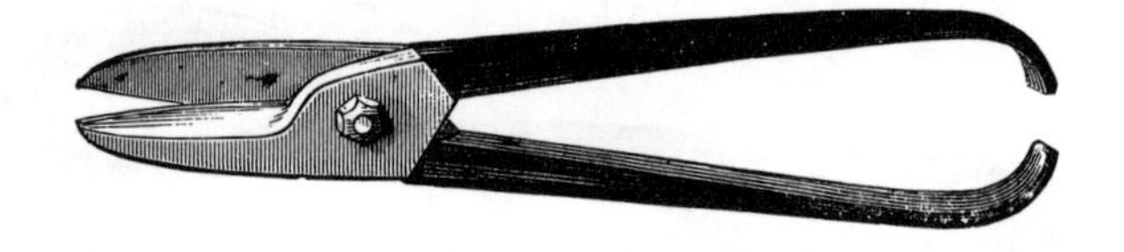

Platers' S[illegible]

No. 1.	Cut 3½ inches, ...	each,	$2 25
No. 2	Cut 3 inches, ...	each,	2 00
No. 3.	Cut 2½ inches, ...	each,	1 75
No. 4.	Cut 2 inches, ...	each,	1 50
No. 5.	Cut 1½ inches, ...	each,	1 25
No. 2.	Circular, ...	each,	2 75
No. 3.	Circular, ...	each,	2 50
No. 4.	Circular, ...	each,	2 25

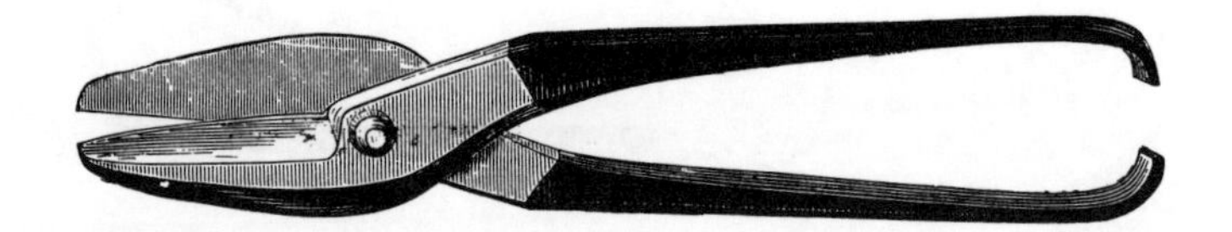

Straight Handle Shears.

We can furnish straight handle shears with circular blades when so ordered.

No. 7.	Hand, cut 4 inches, ...	each,	$2 50
No. 8.	Hand, cut 3½ inches, ...	each,	2 00
No. 9.	Hand, cut 3 inches, ...	each,	1 50
No. 10.	Hand, cut 2½ inches, ...	each,	1 40

COTTON BALE SNIPS.

Cotton Bale Snip.

The above Snip is made expressly for cutting the ties from cotton or other bales.

No. 20. Cotton Bale Snips, length 23 inches, ... each, **$6 00**

THE "LYON" SHEAR.

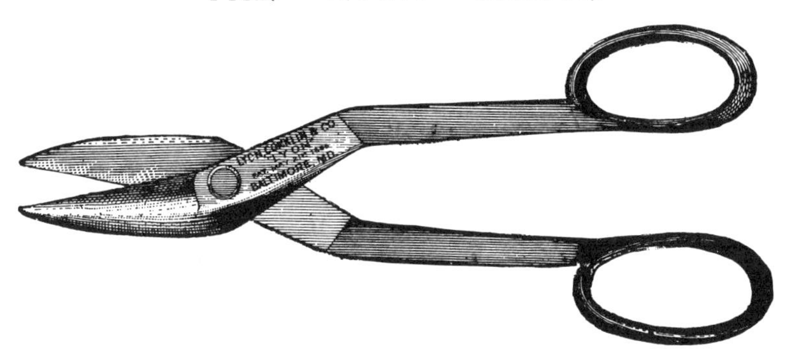

For Cutting Scrolls and Circles.

These Shears are especially adapted to cornice and tin work, and are made so as to easily cut circles, scrolls, etc. The blades are rounding and very sharp pointed, and can be used for the most delicate work. They are made of the best material; have forged handles and steel blades, and are fully warranted.

No. 165.	Hand, cut 4½ inches,	each,	**$3 00**
No. 170.	Hand, cut 4 inches,	each,	**2 50**
No. 180.	Hand, cut 3½ inches,	each,	**2 00**
No. 190.	Hand, cut 3 inches,	each,	**1 50**

COMBINED SNIP AND PIPE CRIMPER.

Double-Cutting Shear and Pipe Crimper.

These double-cutting Shears combined with a Pipe Crimper are now well known. The blade is pointed and readily inserted in the metal at the point desired to begin the cutting. They are adapted to cutting off the bottoms of pails, cans, etc., and suitable for cutting round or square work. The crimping attachment is designed for crimping any kind of sheet metal pipe, round or square. The parts are interchangeable, and the crimping jaws are of steel.

No. 2. Combined Snip and Pipe Crimper, length 13 inches, . . each, **$3 00**

CUTTING NIPPERS.

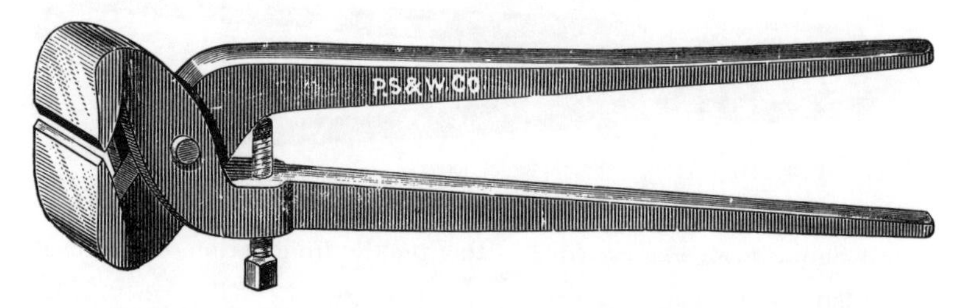

Improved Cutting Nippers.

The jaws of these Nippers are made from best Cast Steel, and each Nipper is tested by cutting steel wire before leaving the factory. After withstanding such a test they are considered perfect and are not warranted against breakage. The long handles give so great a leverage that they are liable to abuse by twisting and prying, and if any such breakage occurs it will be assumed it happens from unfair usage.

No. 0.	Very Large and Strong, 2 inch jaws, 14 inches long, . .	each,	**$3 75**
No. 1.	Extra Large Size, 12 inches long, ...	each,	**2 25**
No. 2.	Large Size, 11 inches long, ...	each,	**2 00**
No. 3.	Common Size, 10 inches long, ...	each,	**1 50**
No. 4.	Small Size, 8 inches long, ...	each,	**1 40**
No. 5.	Small Size, 7 inches long, ...	each,	**1 00**

SHEAR HOLDER.

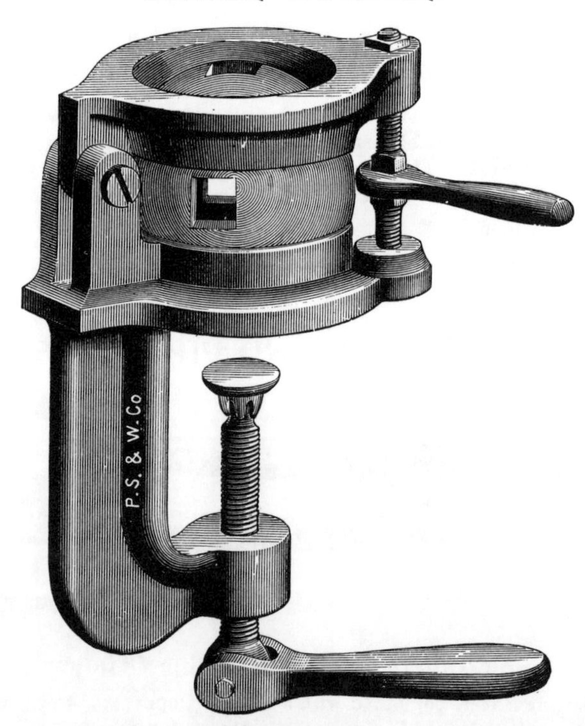

Stow's Improved Shear Holder.

The advantages of this Shear Holder are at once apparent. The stand, clamp and wrench are always attached, the Tinner can use it anywhere on his bench in any part of his shop, and it will hold the bits firmly at any angle desired.

No. 5. Improved Shear Holder, with clamp, weighs 18 lbs., . . each, **$3 00**

STOW'S STAKE HOLDER.

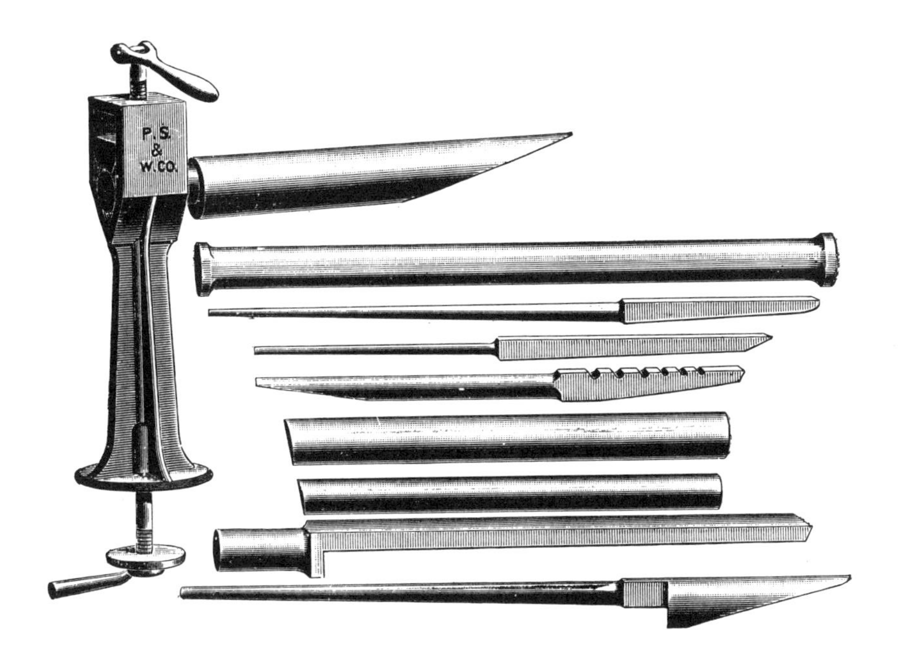

Universal Stake Holder and Stakes.

The illustration above represents our improved Stake Holder and the different tools capable of being used with it.

This holder enables the workman to use the Stakes shown, in any position best suited to the work in hand, without mutilating the bench.

One Stake may be substituted for another with ease.

It is convenient, solid, substantial, and will give satisfaction.

When ordering these Stakes other than in full sets, be careful to mention "for use in Stow's Stake Holder," to avoid confusion with our regular Stakes. The set is made up as follows:

Stake Holder, only,	**$4 00**
Beakhorn for Stake Holder, two pieces,	**10 00**
Blowhorn for Stake Holder,	**3 75**
Creasing and Horn for Stake Holder,	**3 50**
Double Seaming No. 1, for Stake Holder,	**6 00**
Conductor for Stake Holder, two pieces,	**4 00**
Candle Mould for Stake Holder,	**2 00**
Needle Case for Stake Holder,	**1 75**
Full Set as above, complete, with Stake Holder, weighs 139 lbs.,	**$35 00**

TINNERS' STAKES.

WROUGHT IRON, WITH STEEL FACES

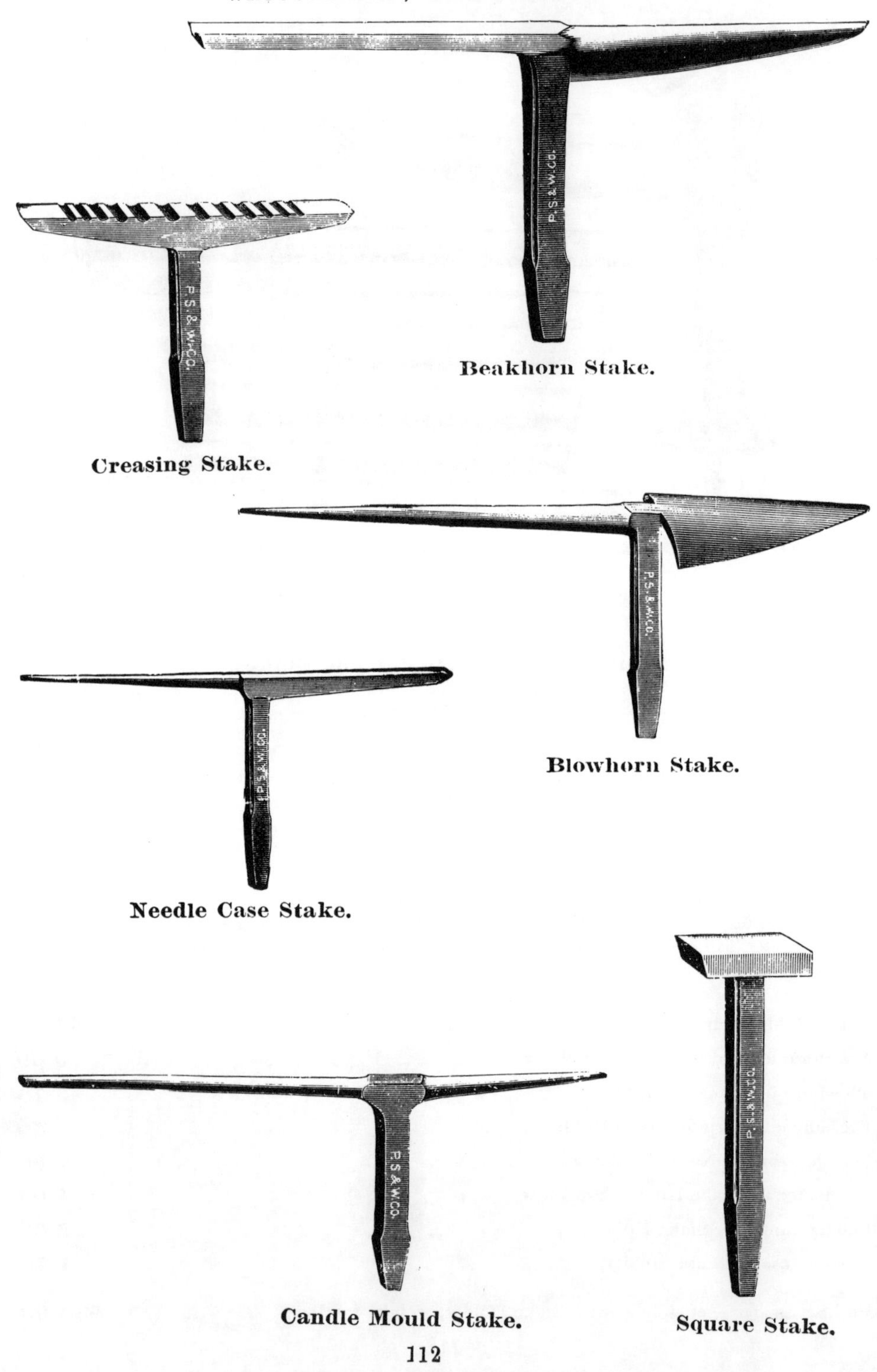

Beakhorn Stake.

Creasing Stake.

Blowhorn Stake.

Needle Case Stake.

Candle Mould Stake.

Square Stake.

PUNCHES.

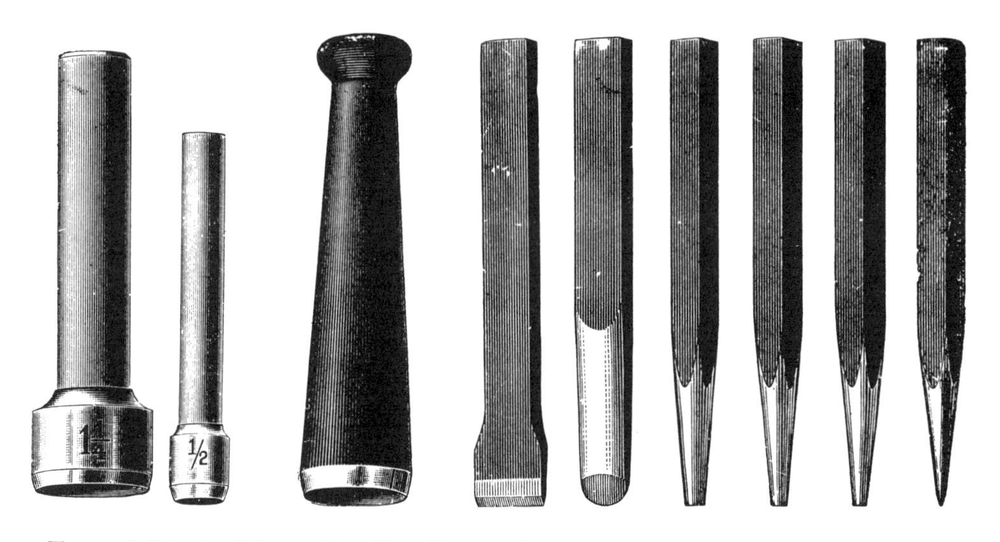

Forged Steel. Wrought Shank. Set Solid Punches, 1-2 Size Cuts.

ARTICLES INCLUDED IN A SET OF TINNERS' TOOLS.

1 Beakhorn Stake, No. 1,	$15 00
1 Blowhorn Stake,	5 00
1 Creasing Stake,	4 00
1 Square Stake,	3 00
1 Candle Mould Stake,	2 75
1 Needle Case Stake,	2 25
1 Set Hollow Punches, one each ½, ¾, 1, 1½, 1¾ inch,	5 50
1 Set Solid Punches, 4 Punches and 2 Chisels,	72
1 Pair Bench Shears, No. 4,	5 00
1 Raising Hammer, each No. 1 and 4, Handled,	3 20
1 Setting Hammer, each No. 2 and 3, Handled,	1 20
1 Riveting Hammer, No. 5, Handled,	38
The above comprise a full set,	$48 00

TINNERS' STAKES.

WROUGHT IRON, WITH STEEL FACES.

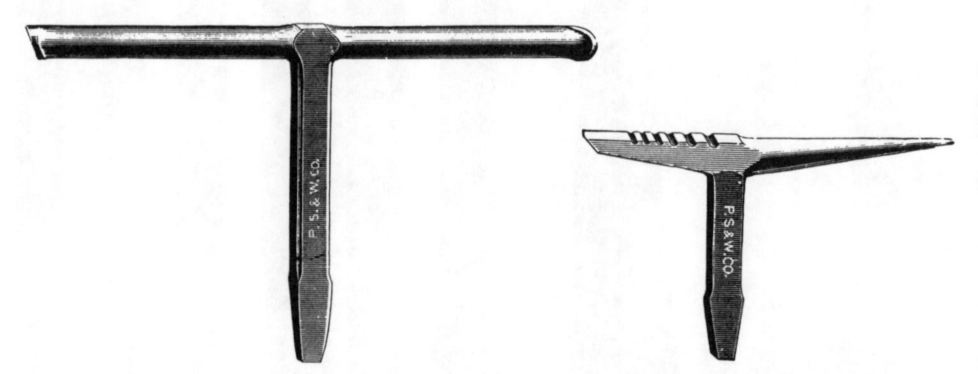

Double Seaming Stake. **Creasing Stake, with Horn.**

Coppersmiths' Square Stake. **Hatchet Stake.** **Bottom Stake.**

Bevel Edge Square Stake.

TINNERS' STAKES.

WROUGHT IRON AND STEEL.

No. 1. Large Stake, or Beakhorn, weighs 45 lbs.,	$15 00
No. 2. Large Stake, or Beakhorn, weighs 40 lbs.,	13 25
No. 4. Large Stake, or Beakhorn, weighs 30 lbs.,	10 00
No. 1. Double Seaming, large end 17 inches, small end 12 inches, weighs 34 lbs.,	9 00
No. 2. Double Seaming, each end 11 inches, weighs 29 lbs.,	8 00
No. 0. Conductor, each end 14 inches, weighs 27 lbs.,	6 00
No. 1. Bevel Edged Square, face 3 x 5 inches, weighs 15 lbs.,	6 00
No. 2. Bevel Edged Square, face 2½ x 4½ inches, weighs 13 lbs.,	5 00
Common Blowhorn, large end 9 inches, small end 17½ inches, weighs 19 lbs.,	5 00
Creasing with Horn, round end 9½ inches, flat end 6½ inches, weighs 12½ lbs.,	4 50
Common Creasing, 14½ inches long, weighs 12½ lbs.,	4 00
Coppersmiths' Square, face 2⅝ x 4½ inches, weighs 11 lbs.,	3 50
Common Square, face 2⅝ x 4½ inches, weighs 11 lbs.,	3 00
Large Square, face 3½ x 5½ inches, weighs 14 lbs.,	7 00
Small Square, face 1½ x 2⅜ inches, weighs 3 lbs.,	2 00
Candle Mould, small end 18 inches, horn 8½ inches, weighs 6½ lbs.,	2 75
Needle Case, flat end 8 inches, small end 10½ inches, weighs 4 lbs.,	2 25
Tea Kettle, with four steel heads, weighs 45 lbs.,	15 75
Steel Heads for Tea Kettle, each,	1 75
No. 1. Hatchet, blade 16 inches long, weighs 14 lbs.,	5 00
No. 2. Hatchet, blade 14½ inches long, weighs 12 lbs.,	4 25
No. 3. Hatchet, blade 13 inches long, weighs 9 lbs.,	3 50
No. 4. Hatchet, blade 11 inches long, weighs 6 lbs.,	2 75
No. 5. Hatchet, blade 9 inches long, weighs 4 lbs.,	2 25
No. 6. Hatchet, blade 7 inches long, weighs 3 lbs.,	1 75
No. 1. Bottom, width 1¾ inches, weighs 4 lbs ,	1 00
No. 2. Bottom, width 1½ inches, weighs 3 lbs.,	80
No. 3. Bottom, width 1¼ inches, weighs 2¼ lbs.,	75
No. 4. Bottom, width 1 inch, weighs 1¾ lbs.,	50

TINNERS' STAKES.

CAST IRON, WITH POLISHED FACES.

Mandrel Stake.

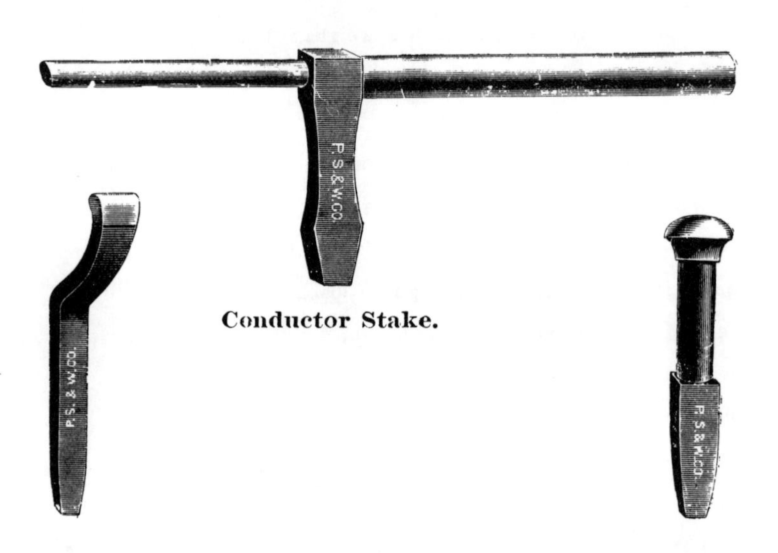

Conductor Stake.

Bath Tub Stake. **Round Head Stake.**

No. 00.	Mandrel, 5 feet long to the Standard, weighs 128 lbs., . .	**$10 00**
No. 0.	Mandrel, 3 feet 4 inches long to the Standard, weighs 87 lbs., . .	**6 00**
No. 1.	Mandrel, 2 feet 10 inches long to the Standard, weighs 44 lbs., .	**5 00**
No. 2.	Mandrel, 2 feet 6 inches long to the Standard, weighs 37 lbs., . .	**4 00**
No. 3.	Mandrel, 2 feet 3 inches long to the Standard, weighs 31 lbs., ,	**3 00**
No. 1.	Conductor, Turned, large end 15 in., small end 11½ in., long, weighs 35 lbs.,	**4 00**
No. 2.	Conductor, Turned, large end 14 in., small end 10 in., long, weighs 25 lbs.,	**3 00**
Round Head, weighs 11 lbs.,	. . . .	**1 25**
Bath Tub, weighs 12 lbs.,	. . . . , ...	**1 25**

TINNERS' STAKES.

WROUGHT AND CAST IRON.

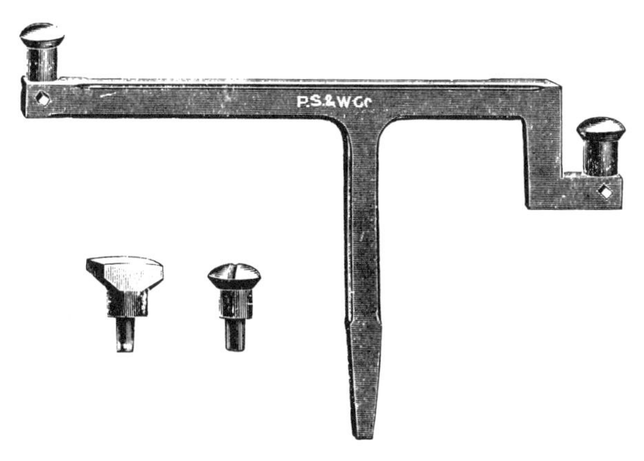

Tea Kettle Stake, with Four Steel Heads.

Hollow Mandrel Stake.

Double Seaming Stake, with Four Heads.

Tea Kettle Stake, Wrought, with 4 Steel Heads, weighs 45 lbs.,		$15 75
Extra Steel Heads, for Tea Kettle Stake,	each	1 75
Double Seaming, with 4 Heads, weighs 85 lbs.,		9 00
Extra Heads for Double Seaming, with 4 Heads,	each,	1 50
Hollow Mandrel, 3 feet 4 inches entire length, weighs 45 lbs.,		5 50
Extra Hollow Mandrel, 4 feet entire length; Round Part, 11¾ inches diameter; Flat Part, 15 inches wide; weighs 300 pounds,		25 00

SWEDGES.

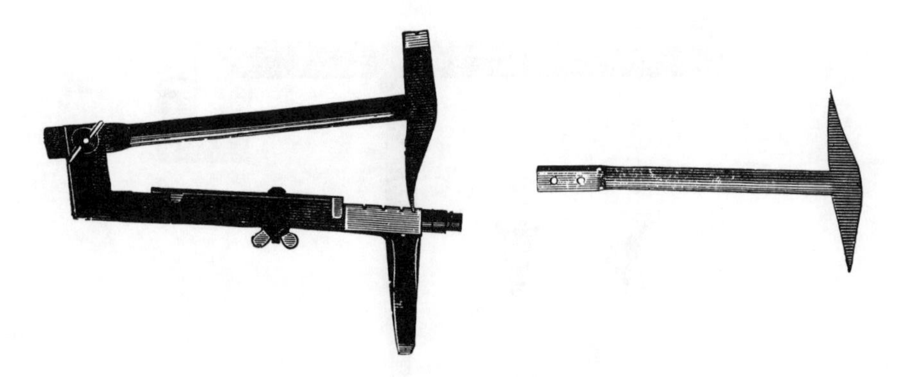

Creasing Swedge.

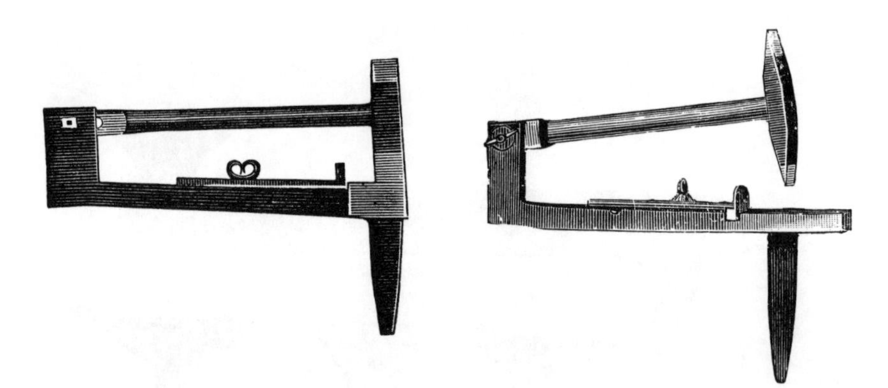

Cullender Swedge. Square Pan Swedge.

Cullender Swedge, ...	$4 75
Square Pan Swedge, ...	5 00
Creasing Swedge, ...	5 25

SQUARE PAN MACHINES.

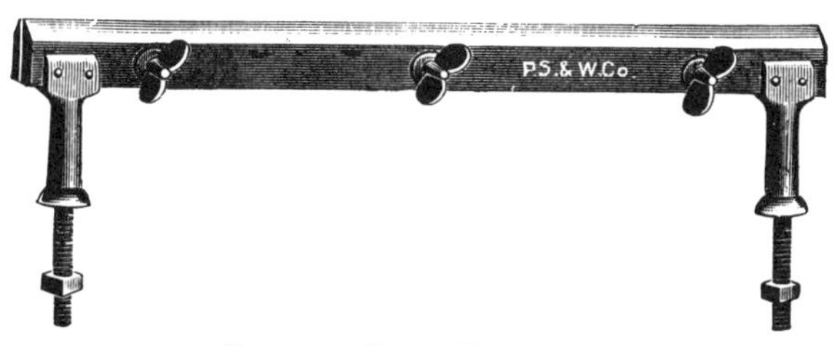

Square Pan Turner.

No. 1.	Square Pan Turner, Steel, 20 inches, weighs 6 lbs., ...	**$2 50**
No. 2.	Square Pan Turner, Steel, 15 inches, weighs 5 lbs., ...	**2 00**

Whitney's Square Pan Former.

Whitney's Square Pan Former, with Steel Blade, 20 inches, weighs 10 lbs., . **$2 50**

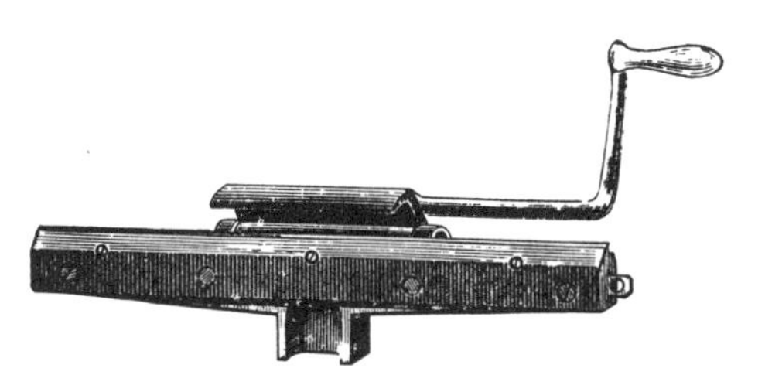

Miller's Square Pan Folder.

This machine is constructed so as to turn the edges of different sizes of square pans, tin or iron, rendering the locks suitable for any size wire.

Miller's Square Pan Folder, Adjustable Gauge, 20 inches, weighs 21 lbs., . . **$5 00**

BENCH PLATES.

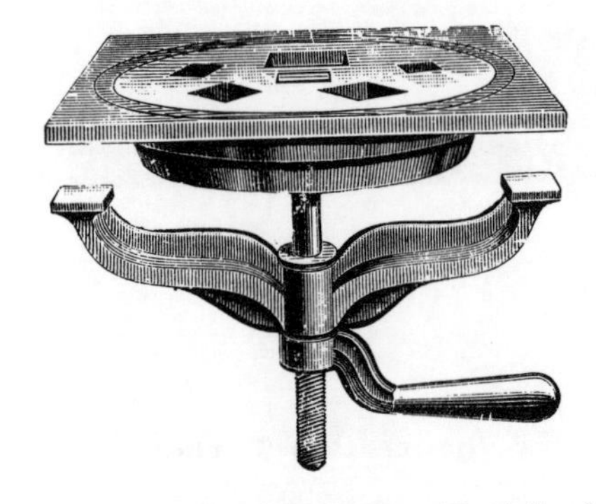

Improved Bench Plate, Polished.

This new and improved Bench Plate is so made that it can be readily inserted in any bench of ordinary thickness, and revolves so that different tools may be used in the same position.

No. 3. 9 x 9 inches, weighs 18 lbs , ... each, **$3 50**

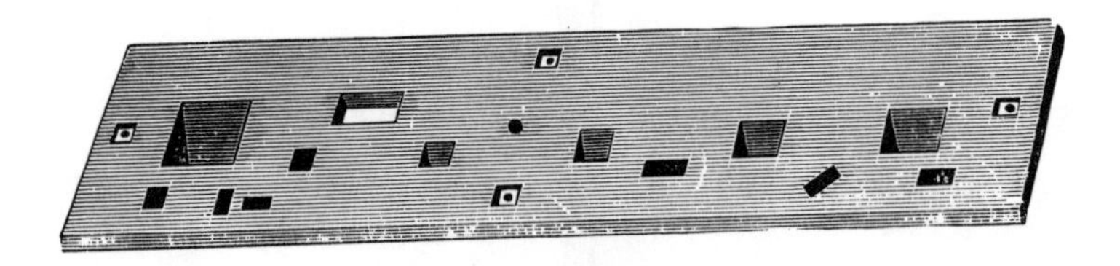

Cast Iron, Polished.

No. 0. 48 x 12 inches, weighs 94 lbs., ... each, **$9 00**

No. 1. 37 x 8 inches, weighs 50 lbs.; . . . , . . each, **5 00**

No. 2. 30 x 8 inches, weighs 37 lbs., ... each, **3 00**

HAMMERS.

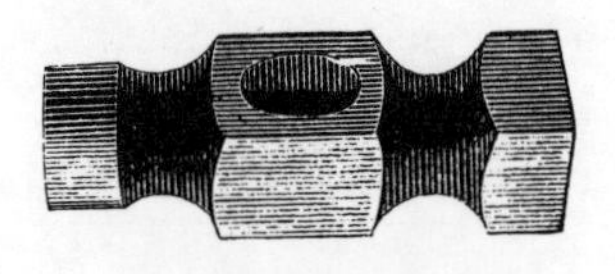

Planishing Hammer.

Planishing Hammers, of different sizes and weights, per pound, **$1 00**

HAMMERS.

Riveting Hammers.

No. 0.	Heavy Work, Bright, 1½ inch, Handled,	per dozen,	$15 50	each,	$1 30
No. 1.	Sheet Iron, Bright, 1⅛ inch, Handled,	per dozen,	9 75	each,	82
No. 2.	Tin, etc., Bright, 1 inch, Handled,	per dozen,	8 31	each,	70
No. 3.	Tin, etc., Bright, ⅞ inch, Handled,	per dozen,	6 75	each,	57
No. 4.	Tin, etc., Bright, ¾ inch, Handled,	per dozen,	5 31	each,	45
No. 5.	Tin, etc., Bright, ⅝ inch, Handled,	per dozen,	4 75	each,	40

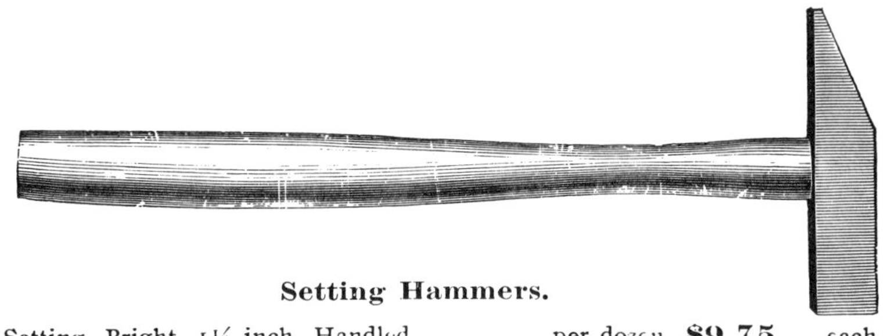

Setting Hammers.

No. 1.	Setting, Bright, 1⅛ inch, Handled,	per dozen,	$9 75	each,	$0 82
No. 2.	Setting, Bright, 1 inch, Handled,	per dozen,	8 31	each,	70
No. 3.	Setting, Bright, ⅞ inch, Handled,	per dozen,	6 75	each,	57
No. 4.	Setting, Bright, ¾ inch, Handled,	per dozen,	5 31	each,	45
No. 5.	Setting, Bright, ⅝ inch, Handled,	per dozen,	4 75	each,	40

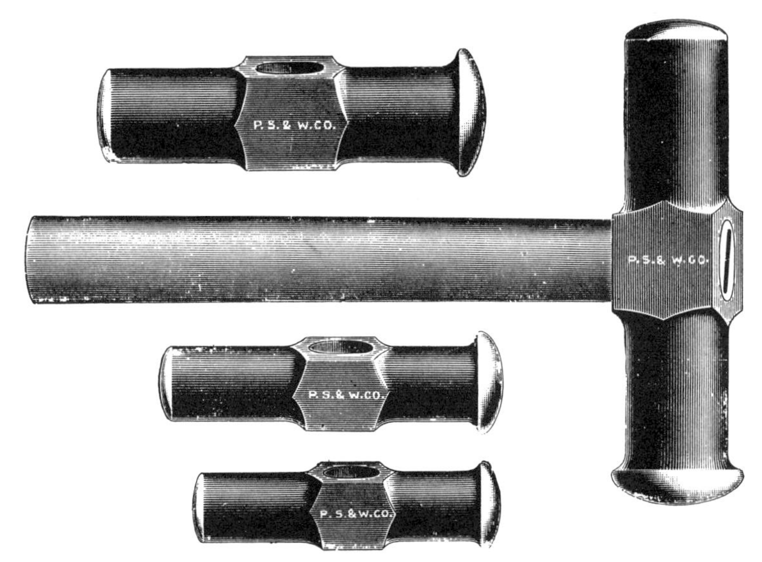

Raising Hammers.

No. 1.	Raising Hammer, weighs 5 lbs.,		each,	$2 25
No. 2.	Raising Hammer, weighs 3¾ lbs.,		each,	1 75
No. 3.	Raising Hammer, weighs 2¼ lbs.,		each,	1 25
No. 4.	Raising Hammer, weighs 1¼ lbs.,		each,	75
Handles,		per dozen	extra,	1 25

HOLLOW PUNCHES.

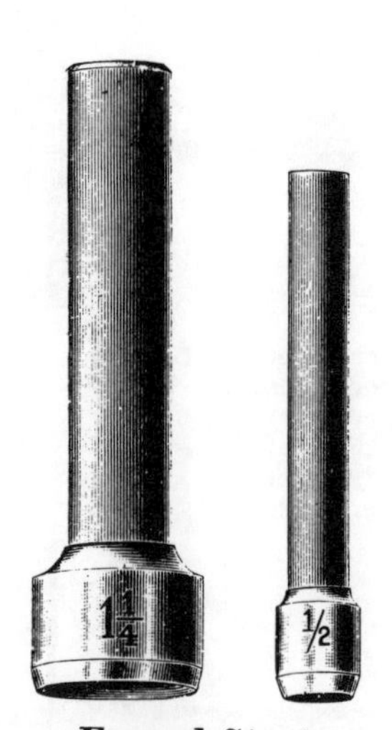

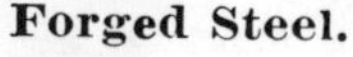
Forged Steel.

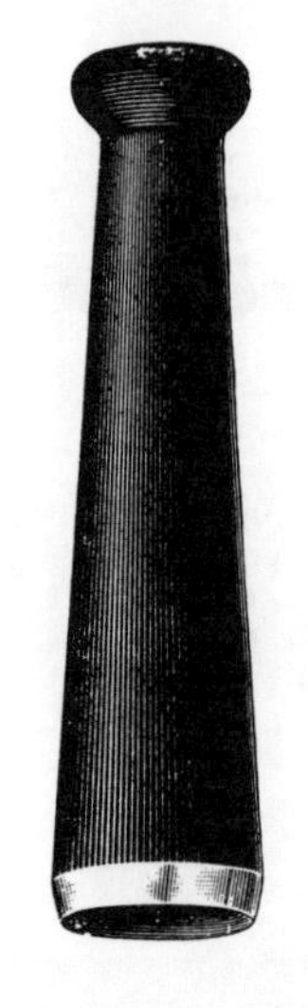

Wrought Shank.

All sizes to and including 1¾ inch diameter, round, ...	per inch,	**$1 00**
All sizes above 1¾ inch diameter, round, ...	per inch,	**1 25**
Oval, ...	per inch,	**1 50**

SPECIAL HOLLOW PUNCHES.

We make all sizes, shapes and kinds for cutting sheet metal, strawboard, paper, etc., Prices for same will be given on receipt of a sample to be cut.

SOLID PUNCHES.

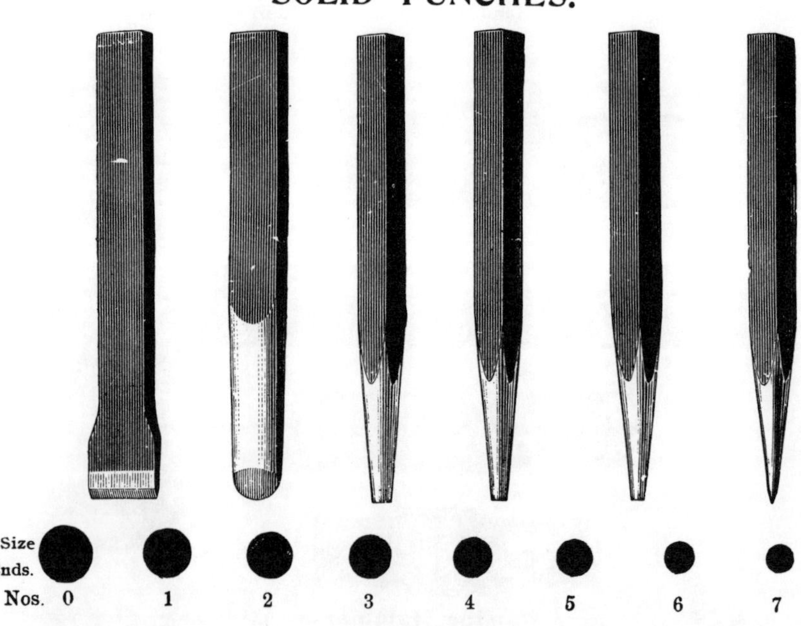

Solid Punches.

Set Solid Punches—4 Punches, 2 Chisels, ...	per set,	**$0 72**
Square, Cast Steel, Nos. 0, 1, 2, 3, 4, 5, 6, 7, 8, and Prick, ...	each,	**12**
Solid Punches, assorted, ...	per dozen,	**1 44**

RIVET SET AND HEADER.

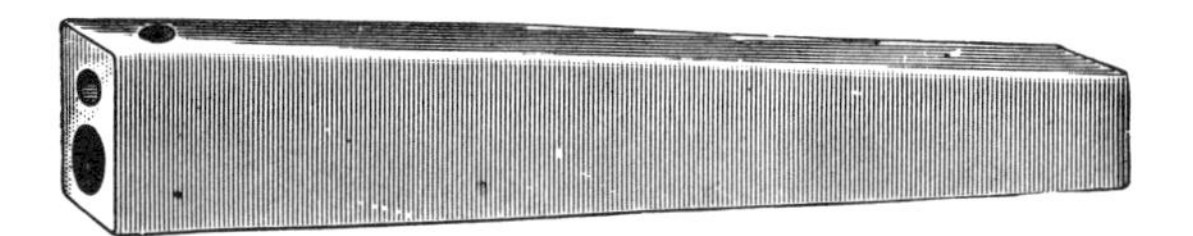

Tinners' Forged Steel Rivet Set, Polished and Blued.

FULL SIZE OF HOLES.

Nos. 00	0	1	2	3	4	5	6	7	8
For Iron Rivets, 14	10 and 12	8	6	4 and 5	2½ and 3	1¾ and 2	1½	1¼ lbs.	10 & 12 oz.
For Copper Rivets, Nos.	5	6	7	8	9	10 and 11	12	13	14

Rivet Sets.

Nos.,	00 and 0.	1 and 2.	3 and 4.	5 and 6.	7 and 8.	Forged steel.	
Each,	**75,**	**63,**	**50,**	**37,**	**32.**	cents,	
Ornament Sets,	. . . .					each,	**$0 38**

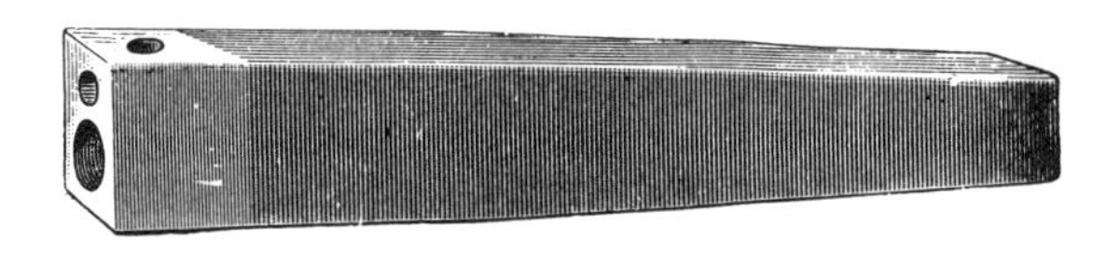

Extra Cast Steel Rivet Set, Japanned, Half Polished.

These goods are precisely the same as those above, except in quality and finish.

Nos.,	00 and 0,	1 and 2,	3 and 4,	5 and 6,	7 and 8,	Cast steel extra.
Each,	**75,**	**63,**	**50,**	**37,**	**32**	cents.

GROOVING TOOLS.

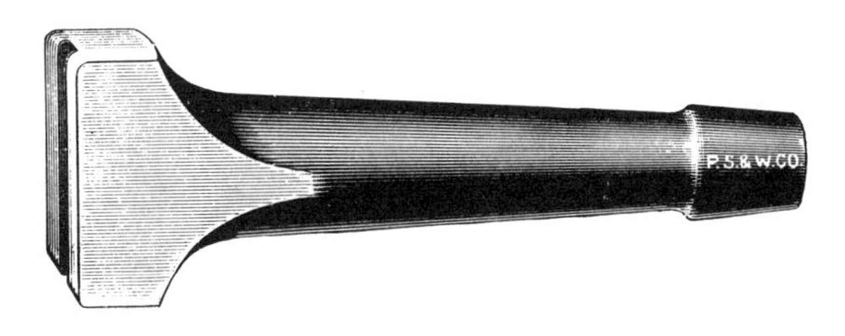

Hand Groover.

Nos.,	00	0	1	2	3	4	5	6	7	8	Forged Steel
Groove,	½	$\frac{7}{16}$	$\frac{13}{32}$	⅜	$\frac{5}{16}$	¼	$\frac{3}{16}$	$\frac{5}{32}$	⅛	$\frac{1}{16}$	inch.
Each,	**75.**	**75,**	**63.**	**63,**	**50,**	**50**	**37,**	**37,**	**25,**	**25**	cents.

CAST STEEL CHISELS.

Lantern Chisel.

Lantern Chisels, common size, ... each, **$0 12**

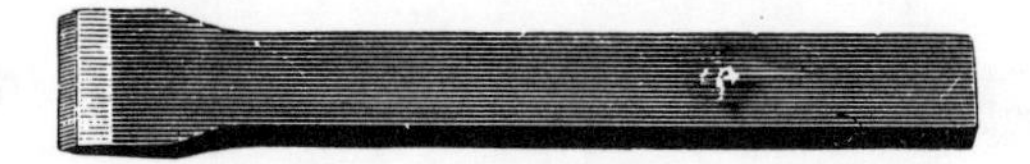

Wire Chisel.

Wire,	¼,	⅜,	½,	⅝,	¾,	⅞,	1,	1⅛,	1¼,	1⅜,	1½,	1¾,	2	inches.
Each,	**8,**	**9,**	**10,**	**11,**	**12,**	**13,**	**14,**	**15,**	**17,**	**19,**	**20,**	**24,**	**29**	cents.

Circular Chisels, ... per inch, **$0 25**

SCRATCH AWLS.

Ring Awl, Solid Steel.

No. 1. Steel Scratch Awls, length, 8½ inches, . . . per dozen, **$1 25**

REESE'S ROOFING TONGS.

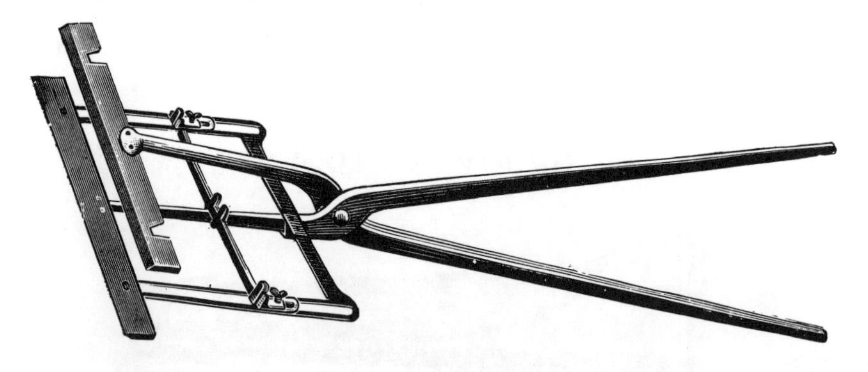

With Adjustable and Removable Gauge.

(Patented Dec. 1884. Improved April 1899.)

These Tongs will turn any edge from ¾ to 10 inches by simply adjusting the thumb screw and sliding the gauge to the desired position. They are adapted to edging and bending all kinds of ordinary sheet metal. They are simple in construction, strong, light and durable. Can be used on No. 22 Iron

Reese's Adjustable Roofing Tongs, ... each, **$6 50**

ROOFING TONGS.

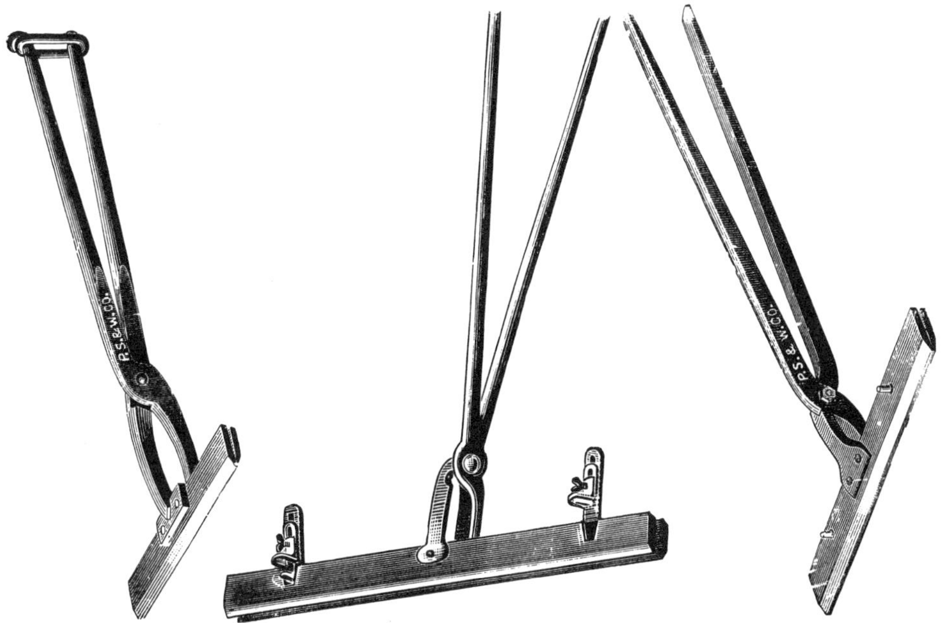

Clamp Tongs. **Improved Tongs.** **Common Tongs.**

The Clamp Tongs are used on the roof for drawing together the two layers of tin before cleating it.

The Improved Roofing Tongs are so made that a single pair may be easily adjusted to turn any size edge from ¾ up to 3 inches by simply adjusting the thumb screws and sliding the gauge to the desired width.

Improved Roofing Tongs, Steel,	per pair	$4 00
Common Roofing Tongs, Steel, Sizes ½, ¾, 1, 1¼, 1½, 1¾ in.,	per set, 2 pairs,	6 00
Clamp Roofing Tongs,	per pair,	2 25

ROOFING TOOLS.

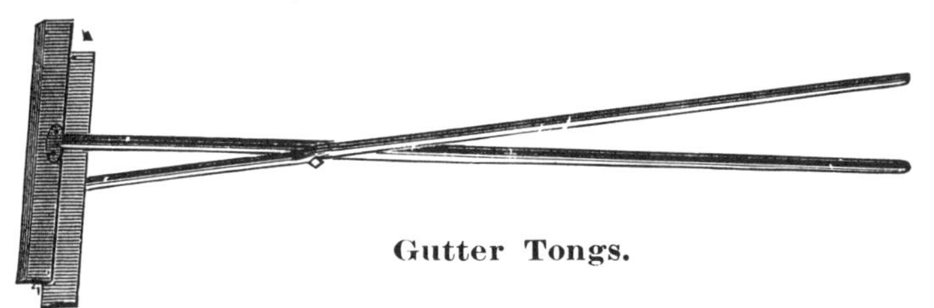

Gutter Tongs.

Roofing Double Seamer. **Improved Wood Roofing Folder.**

Roofing Double Seamers, to match Roofing Tongs,	per set, 2 pairs,	$1 75
Roofing Folder, Common, 30 inch, Wood,		3 50
Roofing Folder, Common, 20 inch, Wood,		2 50
Roofing Folder, Common, 14 inch, Wood,		2 50
Roofing Folder, Improved, 30 inch, Wood, with Gauge,		5 00
Roofing Folder, Improved, 20 inch, Wood, with Gauge,		3 50
Roofing Folder, Adjustable, for 20 or 30 inch, Wood, with Gauge,		6 00
Gutter Tongs,	per pair,	5 00

THE RAPID ROOFING CLEATER AND NAILER.

The simplicity and practicability of this tool and the rapidity with which it can be operated will recommend it at once to all Roofers. With one operation it folds the cleat and nails it to the sheathing, and does it so snugly that the roof will not rattle. Its method of operation is simple. For standing lock roofs the cleat and cleater are hooked over the roofing; a quick movement to one side bends the cleat at right angles and close to the roofing and sheathing. The plunger is then raised and the nail, which has been dropped by the operator—point down—into the funnel shaped pocket on the side of the cleater, slides down against the cleat. A sharp downward stroke of the plunger drives the nail through the cleat and into the sheathing.

It is adapted to tin and sheet metal roofs of various kinds and will nail flat seam roofs rapidly and well.

The use of a hammer is unnecessary and the annoyance and pain of striking the fingers is avoided.

NOTE When ordering please state whether for Common (1 inch) or Wide (1¼ inch) gauge.

The Rapid Cleater and Nailer.

(DANZER'S PATENT FEB. 28, 1899.)

Rapid Roofing Cleater and Nailer, weighs 2¾ lbs., . each, **$2 50**

HEBERLING'S ROOFING DOUBLE SEAMER.

This Roofing Double Seamer is so constructed that but one machine is needed for wide gauge; that is, the gauge made with 1¼ and 1½ inch roofing tongs; and only one is required for common gauge, or 1 and 1¼ inch roofing tongs.

The movable bar, marked A in the illustration, is set in the jaw of the seamer to make the first or single seam. By sliding the movable bar to one side it is released and thrown up against the handles, as shown by the dotted lines in the illustration. The machine is then ready for the second or double seam.

There are two openings in the bottom of this Roofing Double Seamer—one for bending the tin at a slightly acute downward angle, the other for locking or flattening the angle thus made. A down and outward pressure of the handles performs either of these operations.

There is no foot lever of any kind to be operated.

These Double Seamers are very durable and turn an even and perfect lock.

NOTE.—When ordering, state whether for common or wide gauge.

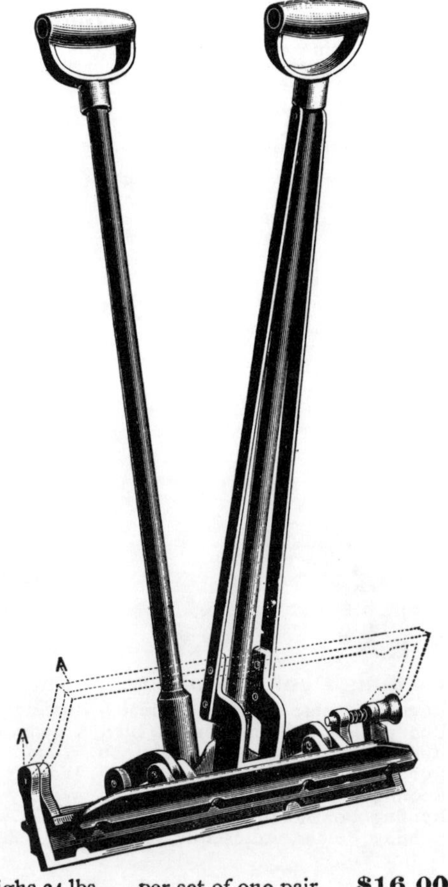

Heberling's Patent, for Standing Lock.

(ONE PAIR CONSTITUTES A COMPLETE SET.)

Heberling's Double Seamer, for Tin Roofing, weighs 24 lbs., per set of one pair, **$16 00**

ROOFING DOUBLE SEAMER.

The cut represents the Double Seamer turning over the edge of the metal. The treadle closes down the edge.

We commend these Roofing Double Seamers for Standing Lock, because the work is done more evenly than by any other process. They turn the metal directly over and under, do not crimp the tin, and leave the formed locks of uniform height. They do not deface the surface of the roof; will work well over uneven roof boarding. Will double seam hips and ridges with perfect ease. They can be used by either hand or foot power. They will do the work much faster, easier and better than by hand and mallet. They are simple in construction, durable and not liable to get out of order. When shipped they are fitted for I. C. Tin.

These Double Seamers are now made and can be furnished adapted to several different styles of Iron Roofing; in fact for the formation of nearly every style of lock now used in applying sheet metals to roofs.

NOTE.—When ordering Seamers please state whether for common or wide gauge.

Common Gauge adapted to **1** x **1¼** inch Tongs.

Wide Gauge adapted to **1¼** x **1½** inch Tongs.

Burritt's Patent, for Standing Lock.

TWO PAIRS CONSTITUTE A SET.

Burritt's Double Seamer for Tin Roofing, . . per set of 2 pairs, **$30 00**

CROSS LOCK SEAMER.

This Machine is constructed for fastening together tin strips for roofing and form the roll ready to lay on the roof. The Machine is noiseless in its working and does not deface the tin. It saves solder, makes smooth roofing, presses the lock tightly together so as to prevent slipping, at the same time leaving the top surface smooth and even for soldering. These Machines are made in two sizes.

Burritt's Patent Cross Lock Seamer.

Burritt's Cross Lock Seamer, 20 inch,	**$44 00**
Burritt's Cross Lock Seamer, 28 inch,	**52 00**

FOUR LEAF CORNICE BRAKE.

Hare's Patent Four Leaf Cornice Brake.

WITH AUTOMATIC FRICTION ROLL CLAMP.

This Brake is warranted to bend No. 20 iron perfectly straight and square its entire length without springing. It is easy and quick to operate. We furnish seven formers with each Machine, to be attached to the Brake for making circular bends as follows:

One each **3, 2½, 2, 1¾, 1½, 1¼, 1** inch.

Clamping jaws of Nos. 36, 42 and 50 open ¾ inch.
Clamping jaws of Nos. 72 and 96 open 1½ inches.
No. 36 has no weight on bending leaf.
Nos. 42 and 50 each has one weight on bending leaf.
Nos. 72 and 96 each has two weights on bending leaf.

It is perfect in all its parts and cannot get out of order. We know it to be superior to any brake ever offered to cornice makers, and we are constantly receiving the most satisfactory testimonials as to its value.

One evidence of its great merit is that it is used in the construction departments of the UNITED STATES GOVERNMENT.

Weights given below are shipping weights.

No. 36	Cornice Brake, clamps 37½ inches, weighs 650 lbs.,	. . . .	**$65 00**
No. 42.	Cornice Brake, clamps 43 inches, weighs 780 lbs.,	. . .	**80 00**
No. 50.	Cornice Brake, clamps 52 inches, weighs 1250 lbs.,	. . . .	**110 00**
No. 72.	Cornice Brake, clamps 74 inches, weighs 2300 lbs.,	. . .	**160 00**
No. 96.	Cornice Brake, clamps 98 inches, weighs 3800 lbs.,	. . . .	**250 00**

EUREKA CORNICE BRAKE.

Hull's Improved Eureka Cornice Brake.

WITH LEVER WEIGHTS AND TREADLE ATTACHMENT.

The demand for a first-class Cornice Brake of a high grade and at a somewhat lower price than our *Celebrated Hare's Four Leaf Cornice Brake* has induced us to put upon the market an eight-foot brake, known as the *Hull's Improved Eureka*, with Lever Weights and Treadle Attachment. This machine is exceedingly well made and simple in construction. We guarantee it to be of equal merit with any brake of its class. This brake is warranted to bend No. 20 Iron perfectly straight its entire length without springing. It is easily and quickly operated.

Seven formers are furnished with each machine for making circular bends.

For each **3, 2½, 2, 1¾, 1½, 1¼** and **1** inch.

The formers are held to the machine by friction clamps and are easily attached or detached. Clamping jaws open 1½ inches.

No. 196. Hull's Improved Eureka Cornice Brake, Clamps 98 ins., weighs 2900 lbs., **$185 00**

PRESSES.

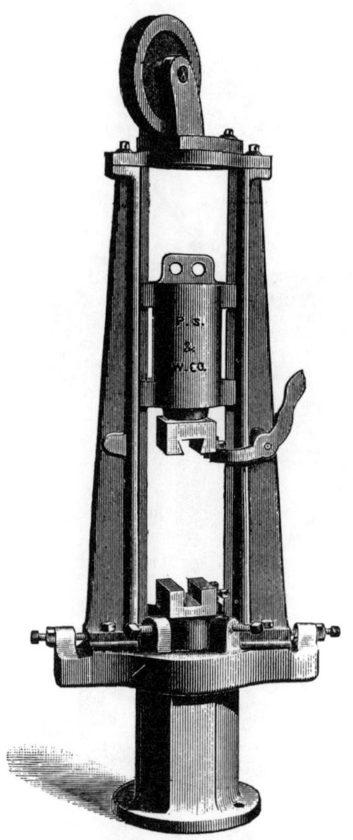

Drop Press.

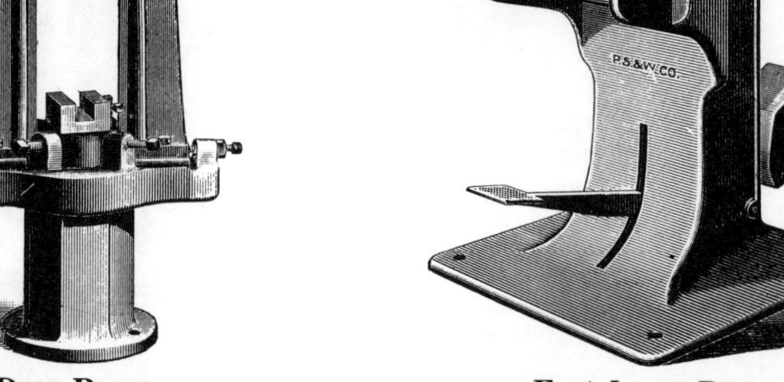

Foot Lever Press.

We manufacture especially for Tinners' Use a light Drop Press, as represented above. It is a convenient press for operating small forming or stamping dies, such as for lettering can covers, etc. As shown it is arranged to work by foot, but can be constructed to work by power. Special attachments can be supplied and prices will be given upon application.

Weight of Bed, 280 lbs.
Weight of Hammer, 70 lbs.
Weight complete, 425 lbs.
Length of Guides, 36 inches.
Width between Guides, 5¾ inches.
Heighth of Bed, 13 inches.
Hammer can be raised 24 inches.
The Guides can be constructed of different lengths if desired.

No. 1. Drop Press, ... **$70 00**

We also manufacture a Foot Lever Press with side extension tables, as illustrated above. It is adapted for cutting tin in many forms which cannot be cut with shears. It has a long slide working in adjustable gibs and has a short stroke, consequently great power in proportion to its weight.

Size of Opening in Bed, 11 x 15 inches.
Stroke of Slide, ¾ inch.
Distance back from centre of Slide, 8½ inches.
Distance from Bed to Slide, when up, 4 to 5½ inches.
Weight Complete, 1300 lbs.

No. 5. Foot Lever Press, ... **$100 00**

MACHINE PARTS.

We give below prices for odd parts of Machines which we are frequently asked to furnish.

Encased and Columbian Machines are the only ones for which we can furnish duplicate parts, and should be ordered by the letters or figures in our catalogue, or stamped upon the part itself.

We do, however, sometimes (but always at the risk of the purchaser) undertake to supply parts to other machines, when the old part for which a new part is wanted is sent us.

When ordering new Cog Wheels for Forming Machines, the diameter of the wheel and hole should be given, and the length or thickness of the wheel measured through the hole or centre.

Encased and Brass Mounted Grooving Machines.

Horn,		$4 50
Crank,		50
Brass Ears,	each,	1 00
Spring and Friction Roll,		75
Bolt and Nut for Stand,		25
Bolt and Nut for Roll,		25
Frame for Brass Mounted,		6 50
Top Casting for Brass Mounted,		3 25

Grannis' Grooving Machines.

Latch,		50
Lower Bar for No. 11,		2 75
Lower Bar for No. 12,		3 00
Rollers,	each,	75

Bigelow's Grooving Machines.

Parts B or C,	each,	70
Friction Roll,		50
Spring,		50

Raymond's and No. 1 Machines.

Brass Top Plate,		1 25
Gauge,		1 25
Top Screws,	each,	10
Cranks,	each,	50
Crank Screws,	each,	40

Stove Pipe Crimper and Beader, Nos. 7, 9, 17 and 19.

Part S.	Hexagon Nut for Crank,		15
Part No. 5.	Crimping Rolls,	each,	1 25
Part No. 6.	Ogee Rolls,	each,	1 00
Part No. 7.	Flange Hexagon Nuts,	each,	15
Part No. 11.	Crank Screw,		40
Part No. 13.	Upper Journal Screw,		15
Part No. 14.	Collars to substitute for Ogee Rolls,	each,	30
Part No. 15.	Cast Frame for Lower Journals,	each,	2 00
Part No. 16.	Cast Frame for Upper Journals,	each,	1 00
Part No 17.	Lower Shaft,		75
Part No. 18.	Upper Shaft,		75
Part No. 19	Crank Shaft,		50
Part No. 20.	Crank,		50
Part No. 21.	Gauge,		50
Part No. 22.	Gauge Screw,		15
Part No. 23.	Pinion,		50
Part No. 24.	Intermediate Gear,	each,	35
Part No. 25.	Gear,	each,	25
Part No. 26.	Lower Shaft Spring,		05

MACHINE PARTS.

Stow's Rim Machine.

Gauge,		$1 25

Crimping Machines.

Gauge,		$1 50
Face and Arbor,		3 00

Beading Machines.

Gauges, for Nos. 1, 2 and 3,	each,	$1 00
Gauge Screw,	each,	25
Gears, for Nos. 1, 2 and 3,	per pair,	1 50
Gears, for Nos. 4 and 5,	per pair,	1 20
Pinion,		75
Rods (2),	per set,	50
Loop,		30
Nuts for End of Roll,	each,	$0 25
Yoke,		25
Cranks,	each,	50
Crank Shaft for No. 4,		2 00
Frame and Top for No. 4,		4 00
Journal Boxes,	each,	25
Cap Screws,	each,	15

Moore's Double Seaming Machines.

Thin Steel Wheel,		$0 75
Crank Screw,		40
Yoke,		50
Gauge,		$0 75
Gears,	per pair,	1 20
Spring,		10

Stow's Double Seaming Machines.

Spring,	$0 50
Arm,	2 00
Treadle,	50
Standard,	3 00
Arbor,	$1 00
Upright or Post,	2 00
Bed,	2 50
Roller, for Setting Down,	2 50

Olmsted's Double Seaming Machines.

Part A, Upright,		$2 00
Part B, Disc Holder,		1 00
Part C, Setting Down Roll,		2 50
Part D, Crank Screw,		40
Part E, Lever,		3 00
Arbor,		1 50
Bottom,		3 00
Step,		50
Front Head,		1 00
Brass Top Plate,		$1 25
Spindle,		75
Head,		2 50
Set of Boxes (3),	per set,	50
Bed Plate,		2 00
Pin,		50
Socket for Stand,		1 50
Treadle and Rod,		1 00

Hulbert's Double Seaming Machines.

Part C,	$2 50
Part D,	1 50
Part F,	50
Part G,	1 00
Part I,	2 50
Part J,	75
Part L,	50
Part P,	$1 00
Part Q,	50
Part R,	3 00
Part T,	25
Part V,	1 00
Part W,	1 00

Stove and Tin Pipe Formers.

Cast Gear or Cogs,	per pair,	$0 50
Cut Steel Gear or Cogs,	per pair,	1 00
Thumb Screws (set of 4),	per set,	2 00
Boxes,	per set,	1 00
Cranks,	each,	$0 50
Lever,		35
Cam,		10

Square Box Folders.

Clamp Bar,	$2 50
Front Gauge,	75
Handle,	50

Smith's Oval Handle Former.

Handle and Bolts,	$0 50
Chisel,	75

MACHINE PARTS.

Jones' Oval Handle Former.

Handle and Cog, combined,	$0 50

Van Bramer's Patent Wire Cutter.

Gauge,	$0 50	Bed Plate,	$3 00
Spring,	25	Handle,	25
Cutter,	50	Gauge Plate,	75
Circle (3 sections),	2 25		

Wright's Sheet Iron Folders.

Upper or Lower Hinge Bar,	each,	$3 00	Legs,	per pair,	$3 00
Steel Folding Plate,		4 00	Leg Bolts,	each,	25
Set Screws,	each,	15	Handle,		75
Gauge,		50	Handle Bolts,	each,	25

Stow's Patent Gutter Beaders.

Jaw or Wing,	$2 00
Frame,	1 00
Lever,	1 00

Squaring Shears, 20, 25, and 30 Inch.

Part A, Long Front Gauges,	each,	$0 70	Legs,	each,	$3 00
Part B, Bolt and Nut for Gauge,		15	Turnbuckles,	each,	40
Part C, Short Front Gauges,	each,	25	Bed for 20 inch,		8 00
Part D, Front Arms,	each,	70	Bed for 30 inch,		10 00
Part EE, Side Springs,	each,	75	Gate or Stock, for 20 or 25 inch,	each,	5 00
Part F, Side Castings,	per set,	3 00	Gate or Stock, for 30 inch,	each,	6 00
Part G, Back Gauges,	each,	70	Grinding Blades,	per pair,	1 75
Part H, Back Gauge Adjusters,	each,	1 00	Bolts for Lower Blade,	per set,	50
Part J, Back Arms.	each,	70	Screws for Upper Blade,	per set,	40
Part K, Connecting Rods,	each,	1 00	Back Arms, with Back Gauge,	per set,	3 00
Part L, Treadle,		3 00			

Newton's Circular Shear.

Shaft,	$1 00	Long Flat Spring,	$0 50
Lever,	1 00	Crank,	50
Holder,	50	Treadle and Rod,	1 50
Thimble,	25	Clamp,	50

Flander's Circular Shear.

Gauge with Rod,		$0 75
Bolts for Cutter Stocks,	per pair,	50
Plug,		50

Savage's Circular Shear.

Holder and Wheel for Burring Attachment,		$1 25
Burring Wheels,	per pair,	1 50
Stock for Small Burr Wheel,		50

Waugh's Rotary Shears, Nos. 1 and 2.

Cutters and Arbors,	per pair,	$5 00	Crank,	$0 50
Bed,		6 00	Slitting and Circle Gauge,	50
Gears,	per pair,	1 25	Sliding Gauge,	25
Upper Box,		25	Back Gauge,	25
Head,		4 00	Gauge Screw,	50
Arm,		3 00		

Notching Machine.

Set of Dies (3 Dies and 1 Punch),	per set,	$5 00
Side Arms,	each,	75

Bolts and Nuts for Bench Shears,	each,	$0 20
Bolts and Nuts for Snip Shears,	each,	12
Hollow Mandrel Fasteners,	each,	50

INTERCHANGEABLE PARTS

FOR STOW'S ADJUSTABLE BAR FOLDERS, Nos. 52, 152, 54 and 154.

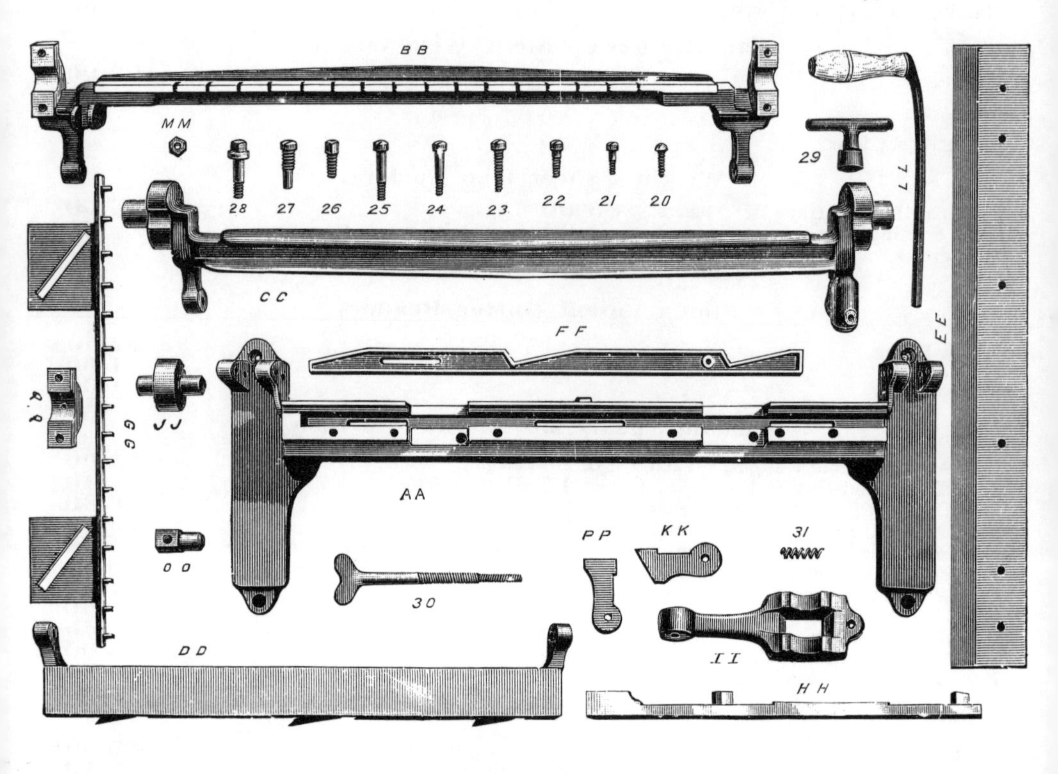

For Nos. 52 and 152 Folders.

AA.	Frame,	$6 00	II.	Shoe,	$0 50
BB.	Jaw,	4 00	JJ.	Friction Roller,	50
CC.	Bar,	4 00	KK.	Stop,	20
DD.	Wing for Bar,	3 00	LL.	Handle,	40
EE.	Blade,	4 00	MM.	Set Nut for Screw No. 23,	15
FF.	Wedge,	50	OO.	Nut for Gauge Screw,	75
GG.	Gauge,	3 50	PP.	Stop,	20
HH.	Slide,	1 00	QQ.	Cap,	50

Screws.

No. 20.	Wedge Screw,	$0 15	No. 26.	Handle Gauge Set Screw,	$0 10
No. 21.	Stop Screw,	15	No. 27.	Wing Screw,	15
No. 22.	Cap Screw,	15	No. 28.	Wedge Screw,	20
No. 23.	Shoe Set Screw,	15	No. 29.	Wrench,	10
No. 24.	Blade Screw,	15	No. 30.	Gauge Screw,	50
No. 25.	Frame Screw,	15	No. 31.	Gauge Springs, per pair,	10

For Nos. 54 and 154 Folders.

AA.	Frame,	$7 00	DD.	Wing for Bar,	$4 00
BB.	Jaw,	6 00	EE.	Blade,	5 00
CC.	Bar,	5 00	GG.	Gauge,	3 50

Other parts same as No. 52 and 152 Folders.

INTERCHANGEABLE PARTS

For Stow's Encased and Columbian Wiring Machines.

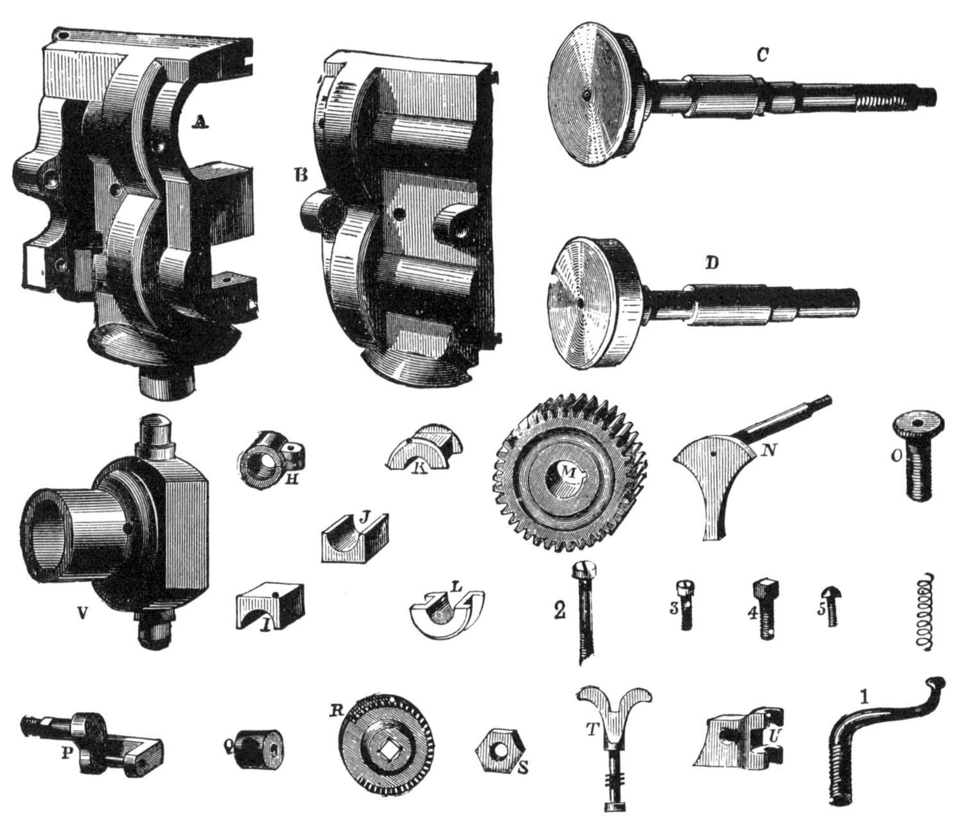

For Encased and Columbian Wiring Machines.

A.	Frame,	$2 50
B.	Cap,	1 50
C.	Upper Roller,	2 50
D.	Lower Roller,	2 50
H.	Clasp Nut,	60
I.	Front Upper Box, for Upper Roller,	25
J.	Front Lower Box, for Upper Roller,	25
K.	Front and Back, Top Boxes for Lower Roller,	25
L.	Front and Back, Lower Boxes for Lower Roller,	25
M.	Gear,	60
N.	Sliding Gauge,	1 25
O.	Sliding Gauge Nut,	75
P.	Forming Gauge for Wiring Machine,	75
Q.	Forming Gauge Roller for Wiring Machine,	25
R.	Forming Gauge Worm Gear, for Wiring Machine,	1 00
S.	Forming Gauge Nut for Wiring Machine,	$0 25
T.	Worm Gear Screw,	75
U.	Worm Gear Screw Holder,	25
V.	Rocking Box,	1 50
No. 1.	Crank Screw,	40
No. 2.	Cap Screw,	15
No. 3.	Lower Box Screw,	10
No. 4.	Worm Gear Screw Holder Bolt,	15
No. 5.	Clasp Nut Screw,	15
	Spring,	10
	Set Nut for Part O,	10
	Faces only for Setting Down Machine,	1 25
	Gears, for Faces,	50
	Shaft, for Faces,	1 25
	Top Plate for Setting Down Machine,	1 25
	Lower Post or Stand for Encased Setting Down Machine,	2 00

The corresponding parts of other Machines in the set bear the same letters or figures as far down as the letter O; also Rocking Box V. Those bearing the letters from P to U, inclusive, are found only in the Wiring Machine.

The above prices apply to parts of all Encased and Columbian Machines.

WIRE GAUGES.

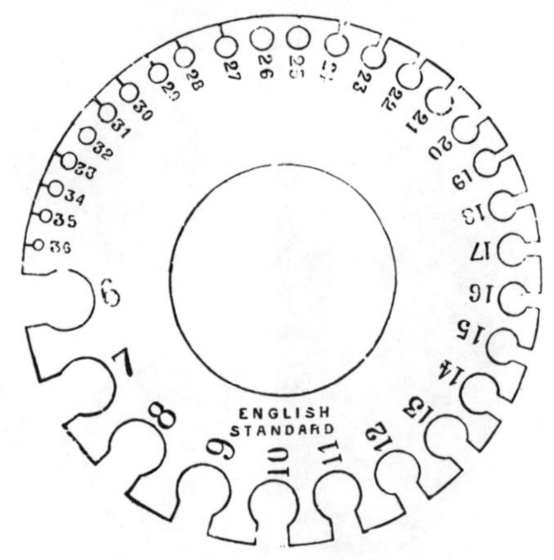

Round, Polished Steel.

ENGLISH STANDARD, CAST STEEL.

No. 1.	Round, Large, Nos. 0 to 36, ...	per dozen,	**$24 00**
No. 2.	Round, Small, Nos. 6 to 36, ...	per dozen,	**15 00**
No. 11.	Round, Large, Nos. 0 to 36, Stub's Pattern, Tapered, . . .	per dozen,	**36 00**

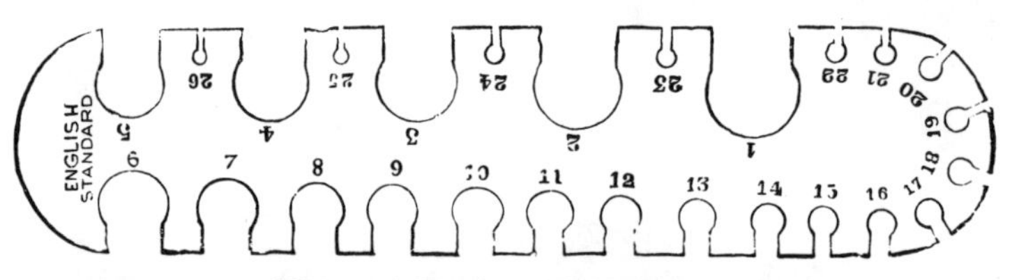

Oblong, Polished Steel.

ENGLISH STANDARD, CAST STEEL

No. 3.	Oblong, Large, Nos. 0 to 36, ...	per dozen,	**$24 00**
No. 4.	Oblong, Small, Nos. 1 to 26, ...	per dozen,	**17 00**

TINNERS' STEEL RULE.

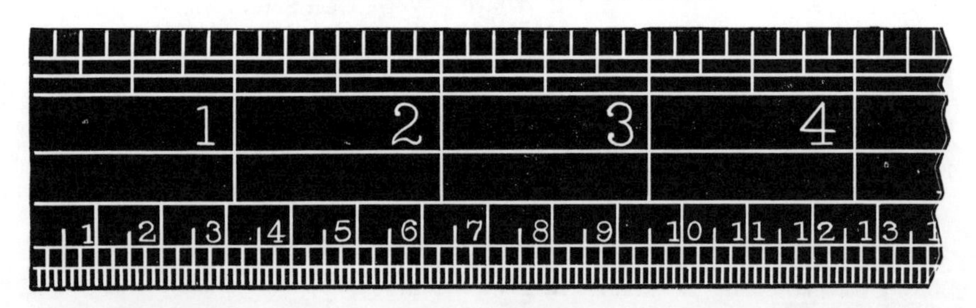

Section of Tinners' Rule. Entire Length, One Yard.

This Tinners' Rule is an invaluable article for any practical tinner, and the cut is an exact representation so far as shown; its entire length is 36 inches. The upper line is the ordinary rule, graduated by eighths of an inch. The lower line shows at a glance the exact circumference of any cylinder by simply ascertaining the diameter, *i. e.*, a vessel 5 inches in diameter the rule indicates to be 15¾ inches in circumference.

The reverse side contains much useful information, in large, plain figures, regarding the sizes of sixty different articles, such as cans, measures, pails, etc., with straight or flaring sides, flat or pitched top, liquid and dry measure in quarts, gallons and bushels. First is given the dimensions for vessels holding 1 to 5 gallons, liquid measure; second, ¼ to 2 bushels, dry measure: third, cans with pitched top, 1 to 10 gallons; fourth, cans with flat top, 1 to 200 gallons; fifth, vessels holding 1 to 8 quarts and ½ bushel to 3 bushels, dry measure.

No. 101.	Tinners' Rule, Polished, ...	each,	**$3 50**
No. 107.	Tinners' Rule, Nickel Plated, ...	each,	**4 00**

SOLDERING COPPER HANDLES.

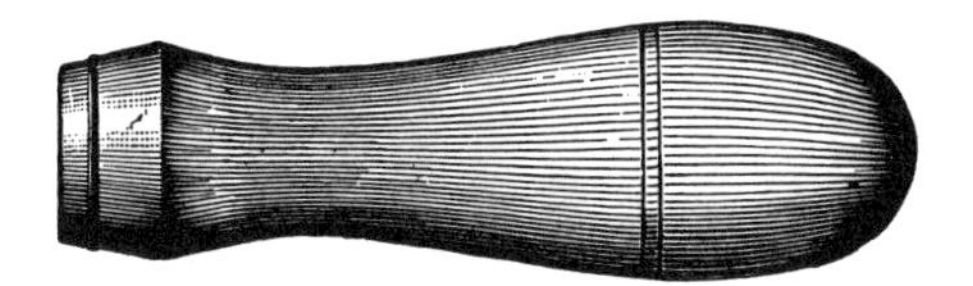

Copper Handle, Wired.

Soldering Copper Handles, ... per dozen, **$0 50**

MALLETS.

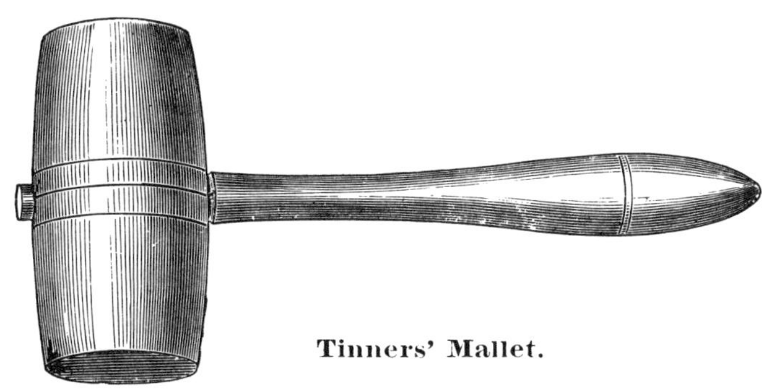

Tinners' Mallet.

Best Seasoned Hickory, assorted, from 2 to 3 inches, . . per dozen, **$1 50**

PLUMBERS' OR MELTING LADLES.

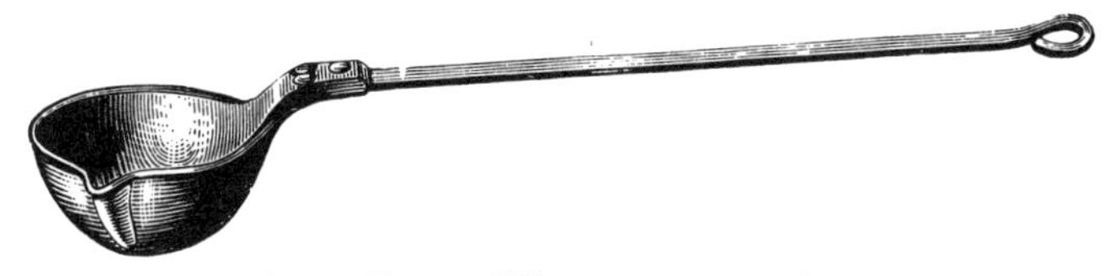

Cast Bowl, Wrought Handle.

Nos.	1	2	3	4	5	6	7
Inches across bowl,	2½	3	3½	4	4½	5	6
Per dozen,	**$1 90**	**2 00**	**2 75**	**3 25**	**4 50**	**5 60**	**7 25**

Long Lip and Long Wrought Handle.

Machinists who have to pour Babbitt metal will at once see the advantage of these Ladles over those in common use.

The nose or lip of the Ladle is made very long, in order that the metal can be poured in places difficult to reach with the ordinary Ladle.

They have a long wrought and wooden handle. We make them in three sizes to hold ¾, 1¼, and 1¾ lbs. of metal.

No. 10.	Diameter, 2 inches.	Will hold ¾ lb.,	per dozen,	**$2 00**
No. 11.	Diameter, 2½ inches.	Will hold 1¼ lbs.,	per dozen,	**2 25**
No. 12.	Diameter, 3 inches.	Will hold 1¾ lbs.,	per dozen,	**2 75**

IMPROVED GAS FURNACE.

Cast Base, Sheet Iron Top.

The above cut represents our Improved Gas Furnace for heating soldering coppers for Plumbers' or Tinners' use. It is light in weight and consumes but little gas. It economizes time, avoids dust and dirt. By regulating the aperture through which the air passes so that the flame has a blue appearance, the very hottest flame produced by gas can be secured.

No. 3. Improved Gas Furnace, ... each, **$1 60**

TINNERS' FIRE POTS.

Cast Iron, Brick Lined. **Sheet Iron, Japanned.**

No. 1 Fire Pot, now so generally introduced, is an universal favorite with Tinners. It is lined with fire brick and made in the most substantial manner. The draft door is in two sections, which economizes fuel.

No. 2 Fire Pot is so constructed that the ashes fall into a pan beneath the coal, and the fire is kept clear and the draft is good. It is light and may easily be carried from place to place at the convenience of the workman.

No. 1. Tinners' Fire Pot, Cast Iron, Brick Lined, ... each, **$3 50**

No. 2. Tinners' Fire Pot, Sheet Iron, Japanned, ... each, **3 00**

SOLDERING COPPERS.

Drawn Copper Bolts, Forged.

WITH SQUARE POINTS FOR COMMON USE. WITH FLAT POINTS FOR BOTTOMS.

HATCHET COPPERS FOR PLUMBERS' USE.

Our Coppers are made of Drawn Copper Bolts of the best quality, and are shaped under a hammer; by this method they are as solid as the metal can be made. They should not be compared with such as are c st from copper ingots.

Nos.	1	1½	2	2½	3	4	5	6	7	8	10	12	14	
Weight,	1	1½	2	2½	3	4	5	6	7	8	10	12	14	pounds per pair

Soldering Coppers, ... per pound, **$0 40**

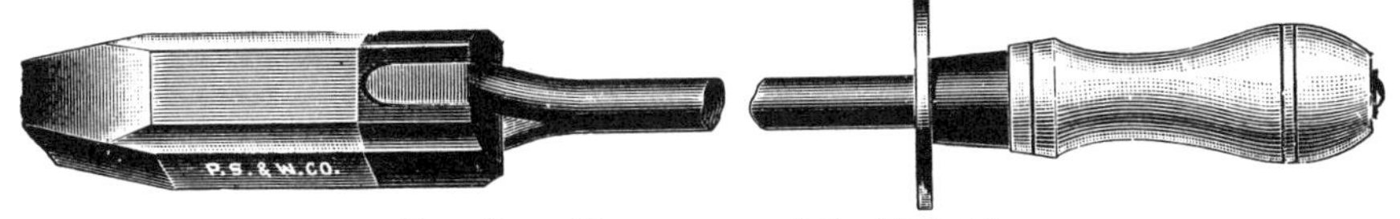

Roofing Coppers, with Shield.

We are making the larger sizes of Soldering Coppers, designed especially for roofers' use, with shield and handle, as represented in the above illustration. They are sold by the pound, the same as regular Soldering Coppers, but at a slightly advanced price.

SOLDERING SETS.

Family Soldering Set.

This little outfit will be found very convenient for household or family use. Each set is complete, and consists of a Soldering Copper, Solder, Rosin and a Scraper, neatly packed in a small wooden box, and are always ready for immediate use.

Family Soldering Sets, ... per dozen sets, **$7 00**

SOLDERING PAN.

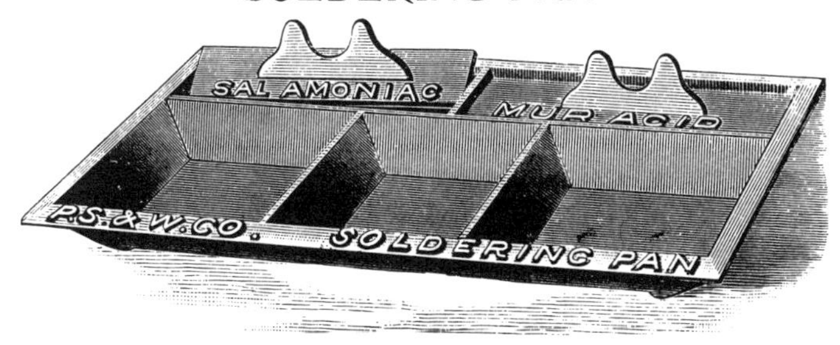

Cast Iron, Plain.

No. 6. Soldering Pan, for Tinners' use, ... per dozen, **$5 00**

IRON KETTLE EARS.

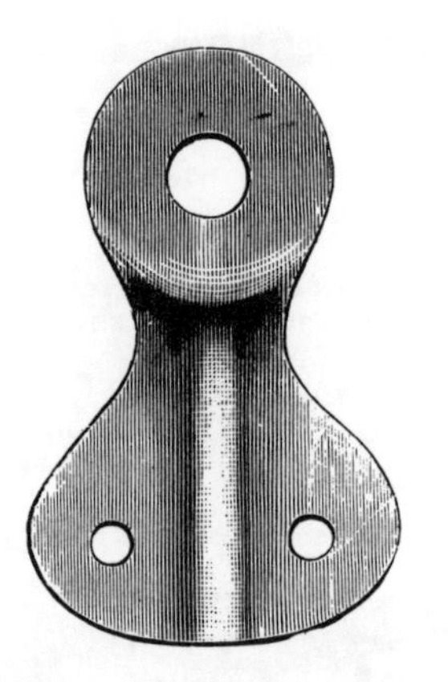

Tinned Kettle Ears—French Pattern.

Tinned Tea Kettle Ears.

No. 1.	Tinned Iron Kettle Ears, Extra Heavy, French Pattern,	per gros., prs.,	$1 00
No. 2.	Tinned Iron Kettle Ears, Extra Heavy, French Pattern,	per gros., prs.,	1 25
No. 3.	Tinned Iron Kettle Ears, Extra Heavy, French Pattern,	per gros., prs.,	1 50
No. 4.	Tinned Iron Kettle Ears, Extra Heavy, French Pattern,	per gros., prs.,	1 75
No. 5.	Tinned Iron Kettle Ears, Extra Heavy, French Pattern,	per gros., prs.,	2 25
No. 6.	Tinned Iron Kettle Ears, Extra Heavy, French Pattern,	per gros., prs.,	2 75
No. 7.	Tinned Iron Kettle Ears, Extra Heavy, French Pattern,	per gros., prs.,	3 50
No. 8.	Tinned Iron Kettle Ears, Extra Heavy, French Pattern,	per gros., prs.,	4 50
No. 4.	Tinned Iron Tea Kettle Ears, ...	per gros., prs.,	1 75
No. 5.	Tinned Iron Tea Kettle Ears, ...	per gros., prs.,	2 10
No. 6.	Tinned Iron Tea Kettle Ears, ...	per gros., prs.,	2 75

MALLEABLE IRON KETTLE EARS.

Full Size Cut of No. 1.

Full Size Cut of No. 2.

Tinned or Black.

No. 1.	Per pound, ...	Black, 15 cents.	Tinned, 18 cents.
No. 2.	Per pound, ...	Black, 15 cents.	Tinned, 18 cents.
No. 3.	Per pound, ...	Black, 15 cents.	Tinned, 18 cents
No. 4.	Per pound, ...	Black, 15 cents.	Tinned, 18 cents.

MALLEABLE EARS AND CLIPS.

Half Size Cut of Clip and Handle, No. 200.

Half Size Cut of Clip, Ear and Handle, No. 100.

Tinned Malleable Ears, Clips and Handles.

The No. 100 combined Ear and Clip is so made that when used with the handle, as shown in the above cut, the handle supports the weight of the vessel when at a right angle with it, and cannot be raised above this point.

The No. 200 Clip, without the Ear, is made in the same manner as the combined Ear and Clip described above.

No. 100.	Patent Combined Ear and Clip, with Handle, Tinned,	per pound,	**$0 18**
No. 200.	Patent Clip, with Handle, Tinned,	per pound,	**18**

MILK CAN HANDLES AND CLIP.

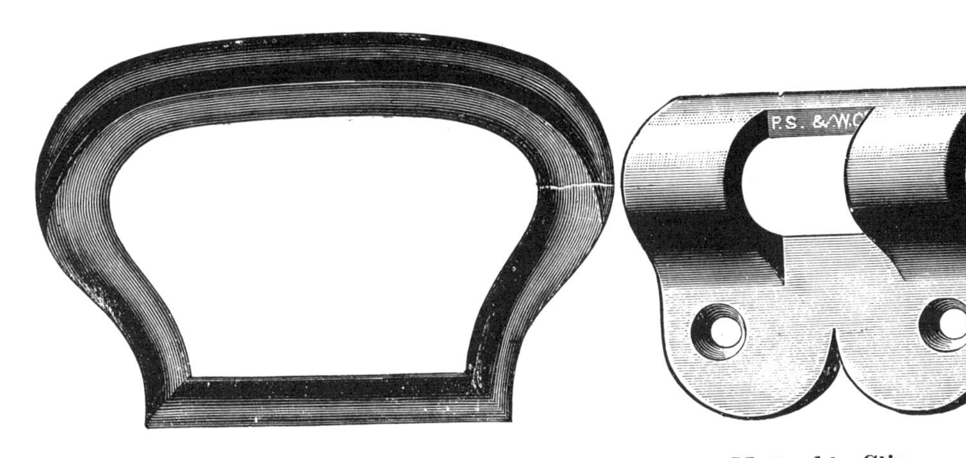

Milk Can or Boiler Handle.

FLUTED AND PLAIN, 3¾ and 4¼ INCHES.

Malleable Clip.

FULL SIZE CUT.

Boiler Handles, Plain,	per pound,	**$0 06**
Boiler Handles, Copper Bronzed,	per pound,	**06**
Boiler Handles, Japanned,	per pound,	**06**
Boiler Handles, Tinned,	per pound,	**15**
Malleable Clips, to fit Boiler Handle, Tinned,	per pound,	**18**

WING DIVIDERS.

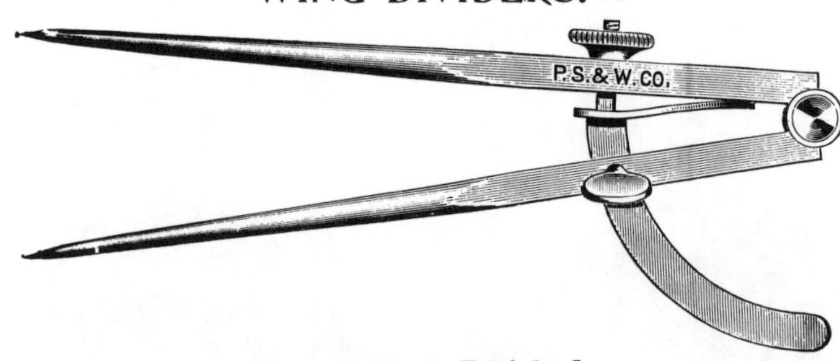

Forged Steel, Polished.

Inches,	5	6	7	8	9	10	12	15	18	24
Per dozen,	$5 50	5 50	6 50	7 50	9 00	10 00	12 00	18 00	25 00	36 00

COMPASSES.

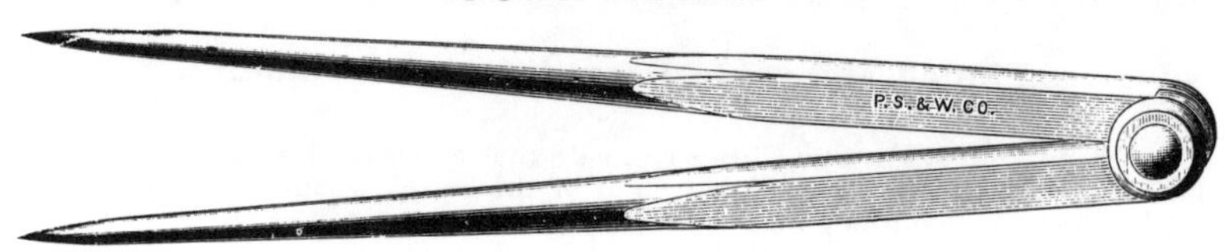

No. 55. Forged Steel, Polished.

Inches,	3	4	5	6	7	8
Per dozen,	$3 25	3 25	3 50	4 00	4 75	5 50

GAS AND BURNER PLIERS.

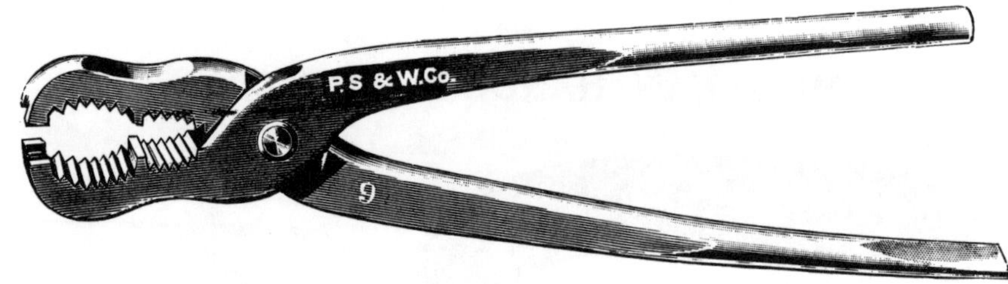

Gas and Burner Pliers, Forged Steel, Polished.

Inches,	5	6	7	8	9	10	11	12	14
Per dozen,	$6 00	7 00	8 00	9 25	10 25	12 50	13 50	14 50	18 00

COMPOUND LEVER CUTTING NIPPERS.

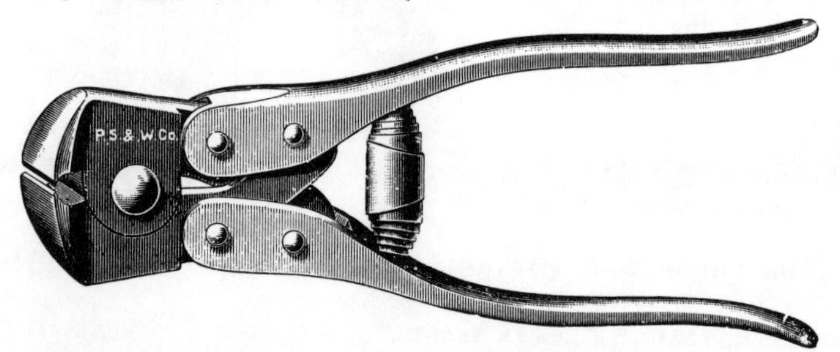

With Detachable Jaws and Volute Steel Spring.

FOR TINNERS' AND GENERAL USE.

No. 7.	7 inches, Polished Faces,	per dozen, $21 00
No. 8.	8 inches, Polished Faces,	per dozen, 24 00

NICKEL PLATED.

No. 7.	7 inches, Nickel Plated,	per dozen, $27 00
No. 8.	8 inches, Nickel Plated,	per dozen, 32 00

FORGED STEEL PLIERS.

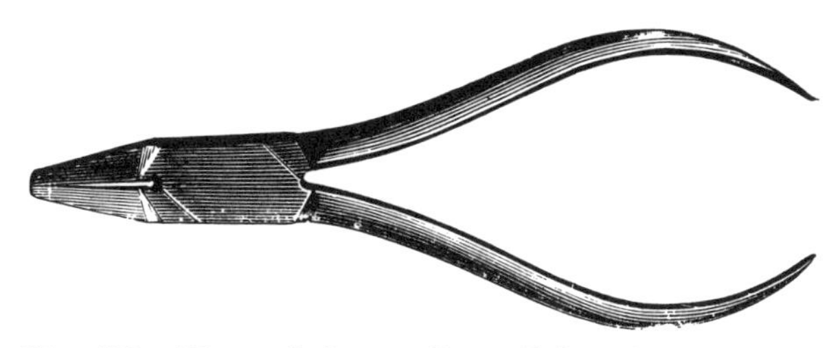

No. 20. Forged Steel, Box Joint, Flat Nose.

Inches,	4	4½	5	6	7	8
Per dozen,	$4 50	4 50	5 00	6 00	8 50	12 50

No. 25. Forged Steel, Box Joint, Round Nose.

Inches,	5	6	7
Per dozen,	$5 00	6 00	8 50

SIDE CUTTING PLIERS.

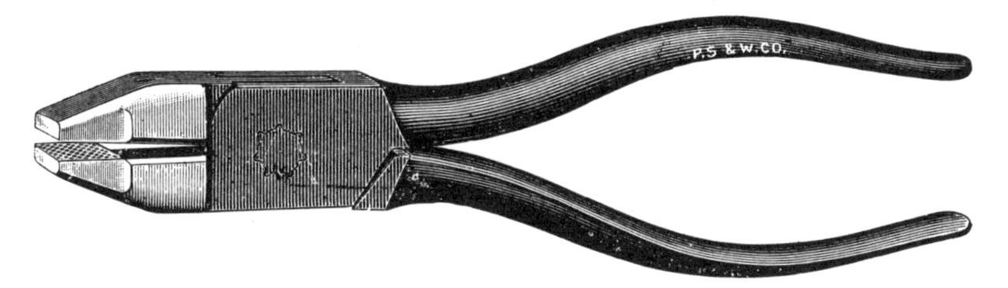

Forged Steel, Box Joint, Raised Cutters.

Inches,	5	6	7	8
Per dozen,	$12 50	13 50	17 00	20 00

NICKEL PLATED.

Inches,	5	6	7	8
Per dozen,	$14 50	15 50	19 50	22 50

COMBINED PLIERS AND CUTTING NIPPERS.

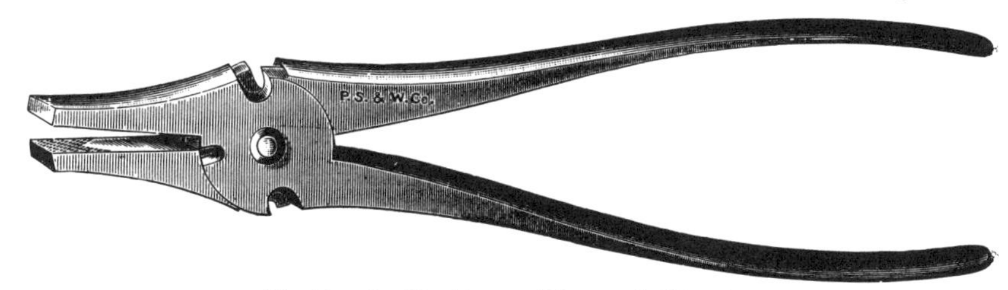

Button's Pattern, Forged Steel.

GUARANTEED OF THE HIGHEST GRADE AND BEST QUALITY.

Inches,	4½	6	8	10	12
Will cut Wire No.	14	11	8	6	5
Per dozen,	$10 00	12 00	15 00	20 00	30 00

GEARED BREAST DRILLS.

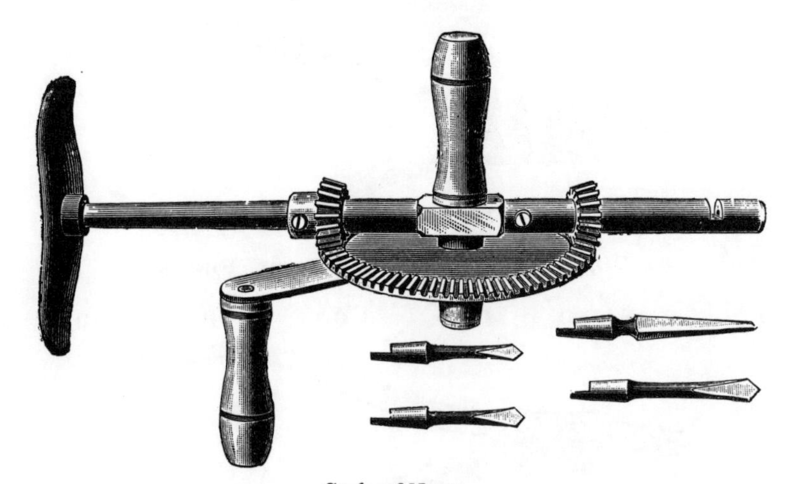

Style of No. 1.

Double Geared, Complete with Drills.

No. 1.	Double Geared Stocks, 4 Drills, Forged Steel, ...	each,	$3 60
No. 2.	Single Geared Stocks, 4 Drills, Forged Steel, ...	each,	3 00
No. 3.	Single Geared Stocks, 4 Drills, Forged Steel, ...	each,	2 50
Extra Drills (3 Drills and 1 Reamer),	. . . . .	per set,	1 25

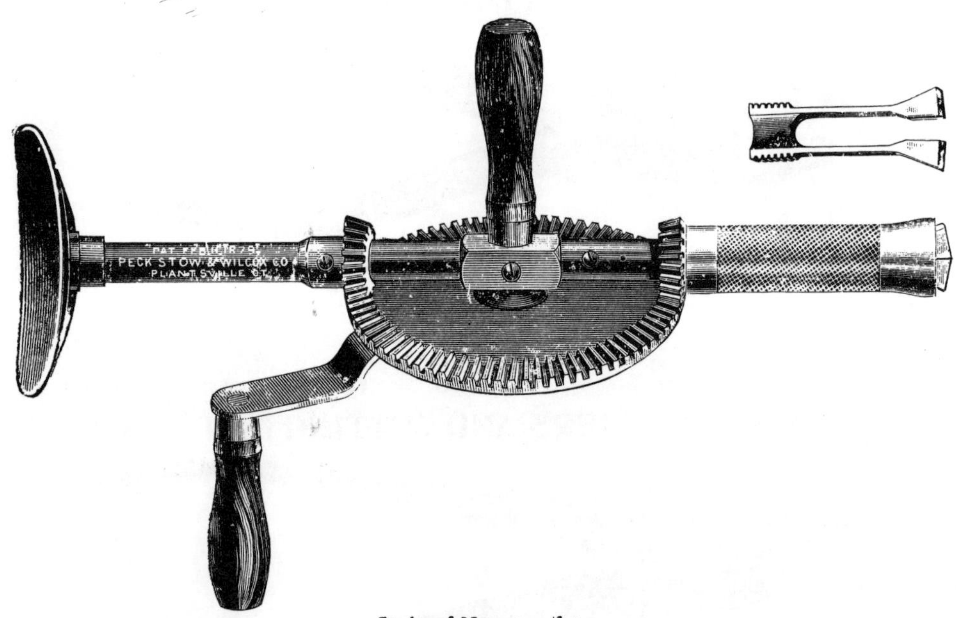

Style of Nos. 4 and 5.

Nickel Plated and Enameled.

WITH ADJUSTABLE CHUCK AND FORGED STEEL JAW.

These Drills are finely finished and perfect in every part. They have Rosewood Handles **and are fitted** with our One Piece, Forged Steel Jaws, which hold equally well round and **square** shank bits.

No. 4.	Nickel Plated, Cut Gear, Extra Finish, ...	each,	$4 50
No. 5.	Enameled, Extra Finish, ...	each,	3 50

RATCHET BIT BRACES.

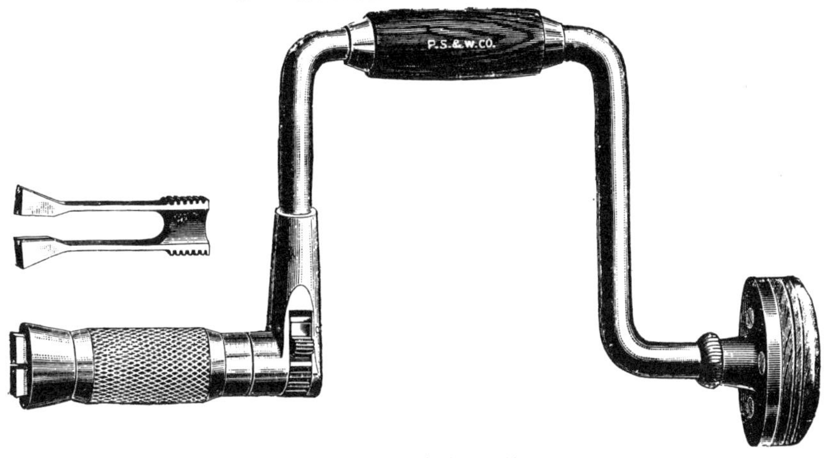

Style of Nos. 1000 to 1004.

Ball Bearing Head, Tempered Forged Steel Jaw.

COCOBOLO HEAD AND CENTRE.

We make a large and complete line of Ratchet and Plain Bit Braces, consisting of over fifty sizes and styles. Illustrations and prices of this entire line will be furnished upon application.

The No. 1000 series represented above is one of the most popular Braces on the market, and we unhesitatingly recommend it as high grade in every respect.

No. 1004.	6 inch, Nickel Plated Sweep,	per dozen,	**$37 50**
No. 1003.	8 inch, Nickel Plated Sweep,	per dozen,	**39 50**
No. 1002.	10 inch, Nickel Plated Sweep,	per dozen,	**45 00**
No. 1001.	12 inch, Nickel Plated Sweep,	per dozen,	**49 00**
No. 1000.	14 inch, Nickel Plated Sweep,	per dozen,	**54 50**

GEARED BREAST DRILL.

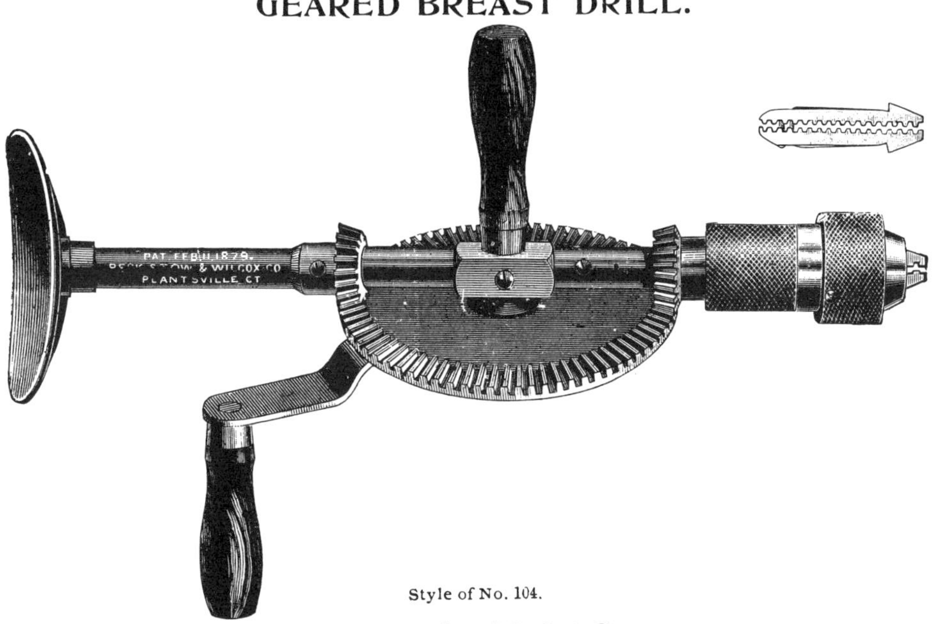

Style of No. 104.

Nickel Plated, with Cut Gear.

WITH SAMSON BALL BEARING CHUCK.

The Ball Bearings in the chuck enforce a stronger grip than can be obtained by any other device. It will hold equally well large expansive bits or the smallest round shank drill. It has Cut Gear, Rosewood Handles and Forged Steel Interlocking Jaws.

No. 104. Nickel Plated, Extra Finish, ... each, **$5 50**

PLUMBERS' SCRAPERS.

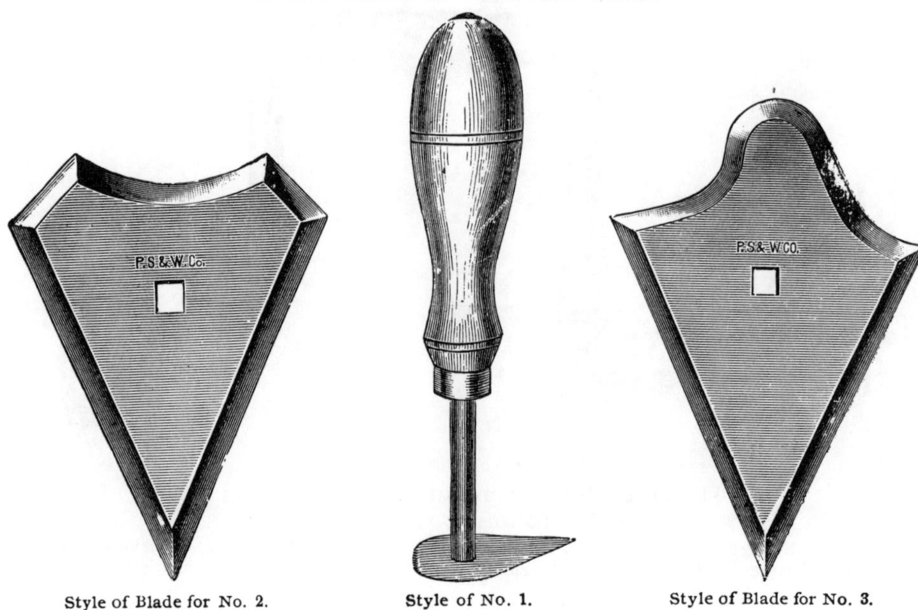

Style of Blade for No. 2. Style of No. 1. Style of Blade for No. 3.

Extra Quality, Forged Steel Blades.

Nos. 2 and 3 are Handled like No. 1; the illustrations above show the shapes of their working edges.

No. 1	Handled, length, 6 inches,	per dozen,	**$3 25**
No. 2	Handled, length, 5½ inches,	per dozen,	**3 25**
No. 3.	Handled, length, 5½ inches,	per dozen,	**3 25**

PARALLEL VISE.

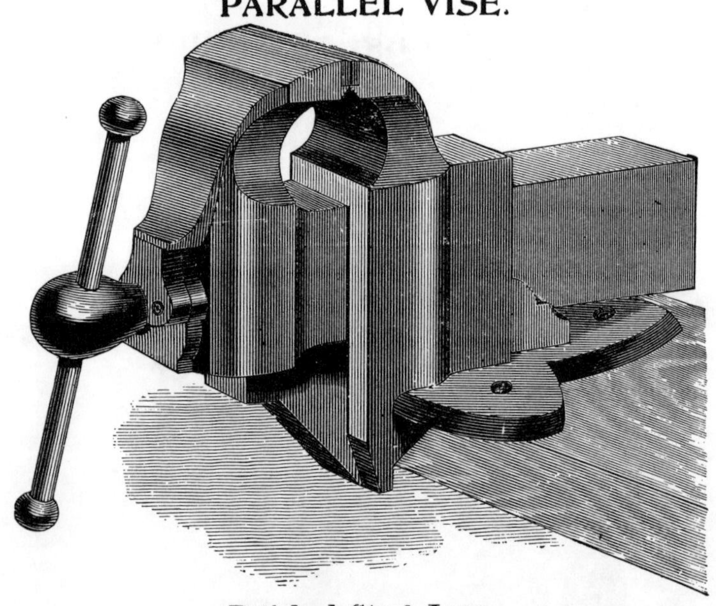

Polished Steel Jaws.

ESPECIALLY DESIGNED FOR TINNERS' USE.

They are made in the most thorough and substantial manner, and are well adapted for general use. The jaws are faced with hardened steel.

No. 100.	Improved Parallel Vise, 3⅝ inch Jaw, weighs 23 lbs.,	each,	**$6 50**
No. 200.	Improved Parallel Vise, 4¼ inch Jaw, weighs 42½ lbs.,	each,	**8 50**

SOLID HANDLE WRENCH.

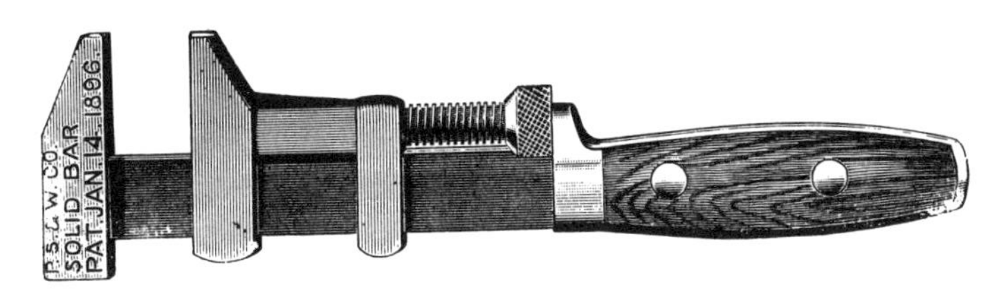

The Solid Handle Wrench.

BY ACTUAL TESTS THE STRONGEST WRENCH MADE.

The Solid Handle Wrench comprises the essential features of a perfect screw wrench—Strength and Simplicity.

The Bar, being of one piece and a solid steel forging, case hardened, has the greatest possible tensile strength and overcomes the weakness and cause of complaint in other screw wrenches, which frequently break at a point on the bar opposite the thumb screw.

The Frame for the wooden handle of the Solid Handle Wrench, being part of the forging of the bar, requires no reduction at any point—as is the case with wrenches having the bar separate from the handle frame. This insures uniform strength throughout.

After a series of tests, we have no hesitancy in warranting the Solid Handle Wrench the strongest and best made.

Packed securely and conveniently in wooden boxes with slide covers.

Inches,	6	8	10	12	15	18	21
Black, per dozen,	$9 00	10 00	12 00	14 00	24 00	30 00	36 00
Bright, per dozen,	10 00	12 00	14 00	16 00	26 00	32 00	38 00

PIPE WRENCHES.

STYLE OF 14 INCH AND SMALLER.

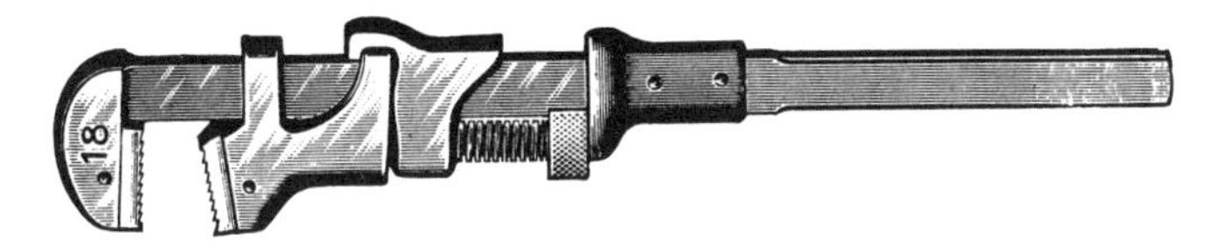

STYLE OF 18 INCH AND LARGER.

Wright's Patent Pipe Wrench.

These Pipe Wrenches have a forged steel bar and head with separate inset jaws easily replaced.

They are very compact and can be operated with one hand. Being narrow back of the jaw they can be used in close quarters. They will not crush the pipe, and the spring is so covered as to prevent its breaking.

Duplicate parts can be furnished.

Length, Inches,	6	8	10	14	18	24
Will open, Inches,	1⅛	1½	1¾	2½	3½	4
Per dozen,	$24 00	24 00	27 00	36 00	48 00	72 00

SCRATCH AWLS.

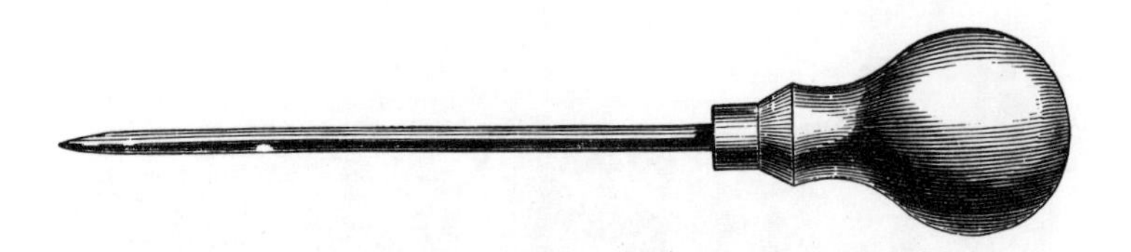

Plain Beech Handles.

No 0.	Plain Beech Handles, Length of Steel, 3 inches, . .	per dozen,	**$0 55**
No. 2.	Plain Beech Handles, Length of Steel, 5 inches, .	per dozen,	**70**

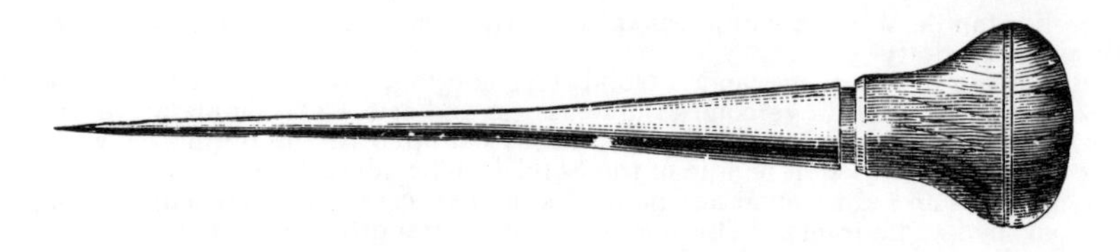

Socket Scratch Awl, Forged Steel.

No. 1. Forged Steel, Polished, Length of Steel, 6 inches, . . per dozen, **$1 70**

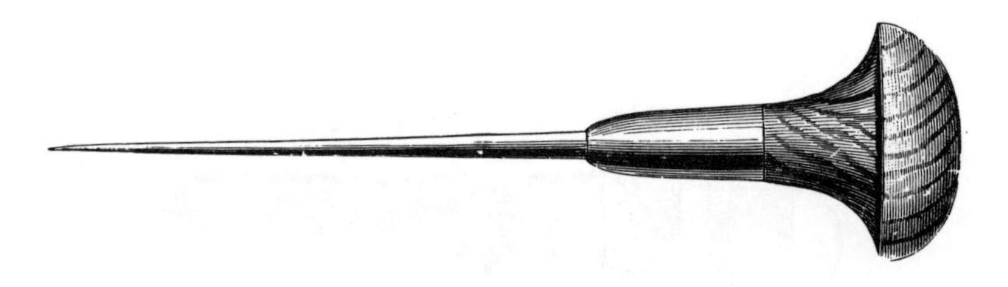

Socket Scratch Awl, Forged Steel.

5 INCH LENGTH OF AWL AND SOCKET.

No. 3. Cherry Handles, with Norway Iron Sockets, . . . per dozen. **$1 70**

FORGED STEEL COLD CHISELS.

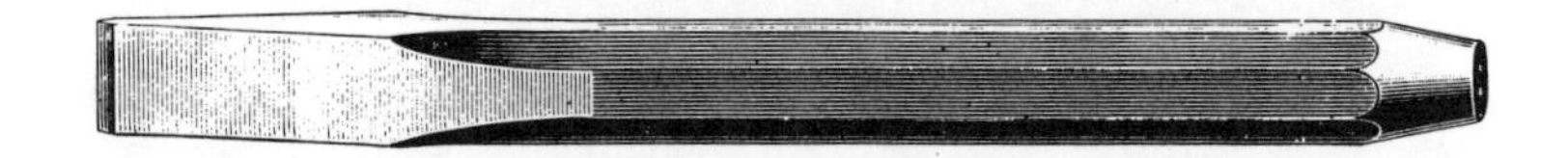

No. 1, Octagon, Half Polished, Turned Heads.

Inches, . .	1/4	3/8	1/2	5/8	3/4	7/8	1
Per dozen,	**$2 00**	**3 60**	**4 50**	**6 00**	**7 50**	**10 00**	**12 80**

TINNERS' STEEL AND IRON SQUARES.

No. 3.	Steel, 2 inches wide, Graduation 1/16, 1/12, 1/8, 1/4 inch, .	per dozen,	**$24 00**
No. 14.	Steel, 2 inches wide, Graduation 1/8, 1/4 inch, . .	per dozen,	**21 00**
No. 4.	Iron, 2 inches wide, Graduation 1/8, 1/4 inch, . .	per dozen,	**12 00**

PASTE JAGGERS.

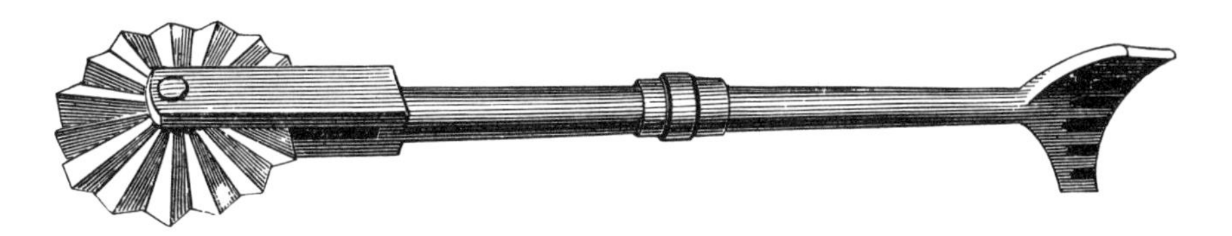

Full Size Cut of Nos. 200 and 300.

Polished and Silver Plated.

No. 200.	Brass, Polished and Lacquered, ...	per dozen,	**$1 50**
No. 300.	Brass, Silver Plated, ...	per dozen,	**1 80**

VENTILATING FLUE STOPPERS.

Smith's Patent.

This safe and ornamental chimney flue stopper we offer to the trade as an article superior to any of the common and often unsafe flue stoppers generally used, and as practically desirable for ventilating rooms. The superiority of ventilating through the chimney is universally acknowledged, and will be apparent to all by simply testing this article.

By the patent attachment of the springs they are easily disconnected, and can readily be changed for flues varying 1½ inches in diameter.

No. 1.	5 to 6 inches, Plain, ...	per gross,	**$18 00**
No. 2.	6 to 7 inches, Plain, ...	per gross,	**21 00**
No. 3.	5 to 6 inches, Painted, assorted colors, ...	per gross,	**24 00**
No. 4.	6 to 7 inches, Painted, assorted colors, . . .	per gross,	**27 00**

THE IDEAL FOOD CUTTER.

The Ideal Food Cutter.

TINNED TO PREVENT RUSTING.

This is unquestionably the best Food Cutter on the market. It will cut rapidly and equally well Meats, Fruits, Vegetables, Crackers, Cheese, etc., and will cut fine or coarse, just as is wanted.

The cutting parts being of Steel gives it pre-eminence over other Food Cutters.

It is exceedingly simple in construction and easily cleaned.

All parts can be duplicated.

Read what two Leaders in the Department of Domestic Science have to say for the Ideal Food Cutter.

Brooklyn, N. Y., Oct. 31, 1898.

The Peck, Stow & Wilcox Co.:
Gentlemen;

Your Ideal Food Cutter has been used in my Schools, both at No. 2 East 42d. Street, New York, and No. 80 Livingston Street, Brooklyn, and I take pleasure in stating it has proved excellent. Wishing you every success,

Very truly yours,

MRS. GESINE LEMCKE.

Canastota, N. Y., Nov. 26th, 1898.

The Peck, Stow & Wilcox Co.:
Dear Sirs;

I received the Ideal Food Cutter while in Auburn and have used it. I find it simple in construction, very easily cleaned, and it does its work perfectly.

Thank you for the opportunity of making its acquaintance.

Yours very truly,

LOUISE A. SCATTERGOOD.

ALBANY SCHOOL OF COOKERY.

No. 25. The Ideal Food Cutter, ... per dozen, **$24 00**